THE COMMONWEALTH AND INTERNATIONAL LIBRARY
Joint Chairmen of the Honorary Editorial Advisory Board
SIR ROBERT ROBINSON, O.M., F.R.S., LONDON
DEAN ATHELSTAN SPILHAUS, MINNESOTA
Publisher: ROBERT MAXWELL, M.C., M.P.

MATHEMATICS DIVISION
General Editors: W. J. LANGFORD, E. A. MAXWELL

A Collection of Problems in Analytical Geometry

PART I

AF874313

A Collection of Problems in Analytical Geometry

PART I

ANALYTICAL GEOMETRY IN THE PLANE

BY

D. V. KLETENIK

Translated by

E. R. DAWSON

Translation edited by

F. J. RAYNER

PERGAMON PRESS

OXFORD · LONDON · EDINBURGH · NEW YORK

TORONTO · PARIS · BRAUNSCHWEIG

Pergamon Press Ltd., Headington Hill Hall, Oxford
4 & 5 Fitzroy Square, London W.1
Pergamon Press (Scotland) Ltd., 2 & 3 Teviot Place, Edinburgh 1
Pergamon Press Inc., 44–01 21st Street, Long Island City, New York 11101
Pergamon of Canada, Ltd., 6 Adelaide Street East, Toronto, Ontario
Pergamon Press S.A.R.L., 24 rue des Écoles, Paris 5e
Vieweg & Sohn GmbH, Burgplatz 1, Braunschweig

Copyright © 1966 Pergamon Press Ltd.
First English edition 1966
Library of Congress Catalog Card No. 66-16466

This book is a translation of the second edition of the Russian volume
СБОРНИК ЗАДАЧ ПО АНАЛИТИЧЕСКОЙ ГЕОМЕТРИИ
(Sbornik zadach po analiticheskoi geometrii)
published by Fizmatgiz, Moscow

This book is sold subject to the condition
that it shall not, by way of trade, be lent,
resold, hired out, or otherwise disposed
of without the publisher's consent,
in any form of binding or cover
other than that in which
it is published.

(2754/66)

CONTENTS

AUTHOR'S PREFACE TO THE FIRST RUSSIAN EDITION

THIS collection of problems is suitable for a course of analytical geometry in those faculties of higher technical institutes which follow the normal syllabus in physico-mathematical and general engineering disciplines. It does not contain any problems in those sections of analytical geometry which are outside the syllabus of higher technical institutes. In compiling the collection the author has paid particular attention to the requirements of theoretical mechanics, since it is this subject which makes the primary and most immediate demands on the subject-matter of analytical geometry.

This collection might also be used in faculties of economics, chemistry, and agriculture, though neither in the presentation nor in the choice of problems are the specific requirements of these faculties taken into account.

The book presents rather more questions on the principal topics of analytical geometry included in the syllabus of higher technical institutes than are usually offered to students for class-work and homework. This fact will enable the supervising teacher to make a choice of the material when using the book, and homework can be given in several alternative forms. Section 17 on the circle is especially comprehensive, and includes a number of questions on the equations of the straight line and of the circle combined; this affords an opportunity of revising the work on one of the most important parts of the course — the equation of a straight line in a plane.

After the theory of second-order curves has been studied it

is useful to give students homework on bringing the general equation of the second degree into canonic form in specific cases; Section 22 contains a sufficient number of questions of similar type but with different numerical data for this purpose.

Having in view the needs of students taking correspondence courses and of those studying higher mathematics on their own, the author has given at the beginning of each section all the fundamental definitions and statements of relevant theorems as well as listing the formulae.

The present collection of problems has been compiled specially for use with N. V. Efimov's *Elementary Course in Analytical Geometry*. The order of presentation of the material and the notation used by Efimov have been followed in compiling this collection.

It is my duty to express my gratitude to the members of the mathematical department of the Moscow Forestry-Engineering Institute for their help in compiling this collection and for their criticism of the first drafts of the manuscript.

D. KLETENIK

AUTHOR'S PREFACE TO THE SECOND RUSSIAN EDITION

FOR its second edition this collection of problems has been considerably revised and increased in content. The most extensive revision has been made in Part I, on "Analytical Geometry in the Plane". Chapter 1, "Elementary Problems in Plane Analytical Geometry", and Chapter 3, "Curves of the First Order", have been radically altered because in the new presentation the concept of a free vector has been excluded from Part I. Chapter 5, which is mainly devoted to illustration of the general theory of second-order curves, has also been considerably altered so as to focus attention on the equations of central curves of the second order, since the simplification of such equations is of great importance in theoretical mechanics. The number of problems in all sections has been increased, from 920 questions in the first edition to 1261 in the second. This increase is largely due to the inclusion of more complicated problems combining different parts of the theory.

The new edition, while maintaining the close connection with N. V. Efimov's *Elementary Course in Analytical Geometry*, is also suitable for use with I. I. Privalov's Course.

I have to thank N. T. Khrolenko for his assistance in helping me to check the answers.

D. KLETENIK

PART I

Analytical Geometry in the Plane

CHAPTER I

ELEMENTARY PROBLEMS IN PLANE ANALYTICAL GEOMETRY

§ 1. The Axis and Intervals on an Axis. Coordinates on a Straight Line

A straight line on which a positive direction has been chosen is called an axis. An interval bounded by any two points A and B on an axis is said to be directed when it is known which point is the beginning, and which point is the end, of the interval. A directed interval beginning at A and ending at B is denoted by the symbol $\overline{AB}$. The measure of a directed interval on an axis is its length taken with the proper sign; the sign is plus if the direction of the interval (i.e., the direction from its beginning to its end) coincides with the positive direction of the axis, and is minus if the direction of the interval is opposite to the positive direction of the axis. The measure of the interval $\overline{AB}$ is denoted by the symbol AB, and its length by the symbol $|AB|$. If the points A and B coincide, the interval defined by them is called a null interval; clearly in this case $AB = BA = 0$ and the direction of such an interval must be considered undefined.

A coordinate system can be set up on any given straight line a by:

1) the choice of a certain interval as the unit of length,
2) the designation of the positive direction on the line a, so

that it becomes an axis (the positive direction on a horizontal axis is generally chosen to be from left to right), and

3) the choice of a particular point O as the origin of coordinates.

The coordinate of any point M of the straight line a (in this coordinate system) is the number x which is equal to the measure of the interval $\overline{OM}$: $x = OM$.

The coordinate of the point O itself is 0.

The symbol $M(x)$ means that the point M has the coordinate x.

If $M_1(x_1)$ and $M_2(x_2)$ are any two points on the straight line a, then the measure of the interval $\overline{M_1M_2}$ is given by

$$M_1M_2 = x_2 - x_1$$

and its length is given by

$$|M_1M_2| = |x_2 - x_1|.$$

1. Construct the points:

$$A(3),\ B(5),\quad C(-1),\quad D(2/3),\quad E(-3/7),\quad F(\sqrt{2}),\quad \text{and}\quad H(-\sqrt{5}).$$

2. Construct the points whose coordinates satisfy the equations:

1) $|x| = 2$; 2) $|x-1| = 3$; 3) $|1-x| = 2$; 4) $|2+x| = 2$.

3. Specify geometrically the position of points whose coordinates satisfy the inequalities:

1) $x > 2$;
2) $x-3 \leqslant 0$;
3) $12-x < 0$;
4) $2x-3 \leqslant 0$;
5) $3x-5 > 0$;
6) $1 < x < 3$;
7) $-2 \leqslant x \leqslant 3$;
8) $(2-x)/(x-1) > 0$;
9) $(2x-1)/(x-2) > 1$;
10) $(2-x)/(x-1) < 0$;
11) $(2x-1)/(x-2) < 1$;
12) $x^2-8x+15 \leqslant 0$;
13) $x^2-8x+15 > 0$;
14) $x^2+x-12 > 0$;
15) $x^2+x-12 \leqslant 0$.

4. Determine the measure AB and the length $|AB|$ of the interval given by the points:

1) $A(3)$, $B(11)$; 2) $A(5)$, $B(2)$; 3) $A(-1)$, $B(3)$;
4) $A(-5)$, $B(-3)$; 5) $A(-1)$, $B(-3)$; 6) $A(-7)$, $B(-5)$.

5. Calculate the coordinate of the point A given:

1) $B(3)$, $AB = 5$; 2) $B(2)$, $AB = -3$;
3) $B(-1)$, $BA = 2$; 4) $B(-5)$, $BA = -3$;
5) $B(0)$, $|AB| = 2$; 6) $B(2)$, $|AB| = 3$;
7) $B(-1)$, $|AB| = 5$; 8) $B(-5)$, $|AB| = 2$.

6. Specify geometrically the position of the points whose coordinates satisfy the following inequalities:

1) $|x| < 1$; 2) $|x| > 2$; 3) $|x| \leqslant 2$:
4) $|x| \geqslant 3$; 5) $|x-2| < 3$; 6) $|x-5| \leqslant 1$;
7) $|x-1| \geqslant 2$; 8) $|x-3| \geqslant 1$; 9) $|x+1| < 3$;
10) $|x+2| > 1$; 11) $|x+5| \leqslant 1$; 12) $|x+1| \geqslant 2$.

7. Determine the ratio $\lambda = AC/CB$ in which the point C divides the interval $\overline{AB}$, given:

1) $A(2)$, $B(6)$, $C(4)$; 2) $A(2)$, $B(4)$, $C(7)$;
3) $A(-1)$, $B(5)$, $C(3)$; 4) $A(1)$, $B(13)$, $C(5)$;
5) $A(5)$, $B(-2)$, $C(-5)$.

8. Given the three points $A(-7)$, $B(-1)$, $C(1)$, determine the ratio λ in which each point divides the interval bounded by the other two points.

9. Determine the ratio $\lambda = M_1M/MM_2$ in which a given point $M(x)$ divides the interval $\overline{M_1M_2}$ bounded by the given points $M_1(x_1)$, $M_2(x_2)$.

10. Find the coordinate x of the point M which lies on the interval $\overline{M_1M_2}$ bounded by the given points $M_1(x_1)$, $M_2(x_2)$ and which divides it in a given ratio λ ($\lambda = M_1M/MM_2$).

11. Find the coordinate x of the mid-point of an interval bounded by two given points $M_1(x_1)$, $M_2(x_2)$.

12. Find the coordinate x of the mid-point of the intervals bounded by each of the following pairs of points:

1) $A(3)$, $B(5)$; 2) $C(-1)$, $D(5)$; 3) $M_1(-1)$, $M_2(-3)$; 4) $P_1(-5)$, $P_2(1)$; 5) $Q_1(3)$, $Q_2(-4)$.

13. Find the coordinate x of the point M given:

1) $M_1(3)$, $M_2(7)$, $\lambda = M_1M/MM_2 = 2$.
2) $A(2)$, $B(-5)$, $\lambda = AM/MB = 3$.
3) $C(-1)$, $D(3)$, $\lambda = CM/MD = 1/2$.
4) $A(-1)$, $B(3)$, $\lambda = AM/MB = -2$.
5) $A(1)$, $B(-3)$, $\lambda = BM/MA = -3$.
6) $A(-2)$, $B(-1)$, $\lambda = BM/MA = -1/2$.

14. Given the two points $A(5)$, $B(-3)$, find the coordinate of the point M which is symmetrical with the point A relative to point B, and of the point N which is symmetrical with the point B relative to point A.

15. The interval bounded by the points $A(-2)$, $B(19)$ is divided into three equal parts. Find the coordinates of the points of division.

16. Find the coordinates of the ends A and B of the interval which is divided into three equal parts by the points $P(-25)$ and $Q(-9)$.

§ 2. Rectangular Cartesian Coordinates in a Plane

A rectangular Cartesian coordinate system is defined when a unit of length for measuring lengths and two mutually perpendicular axes are given and it is known which is the first axis and which the second. The point of intersection of the axes is the coordinate origin and the axes themselves are called the coordinate axes. The first axis is called the x-axis and is denoted by Ox, and the second axis is called the y-axis and is denoted by Oy, O being the coordinate origin.

The coordinates of an arbitrary point M in the given system are the numbers

$$x = OM_x, \quad y = OM_y$$

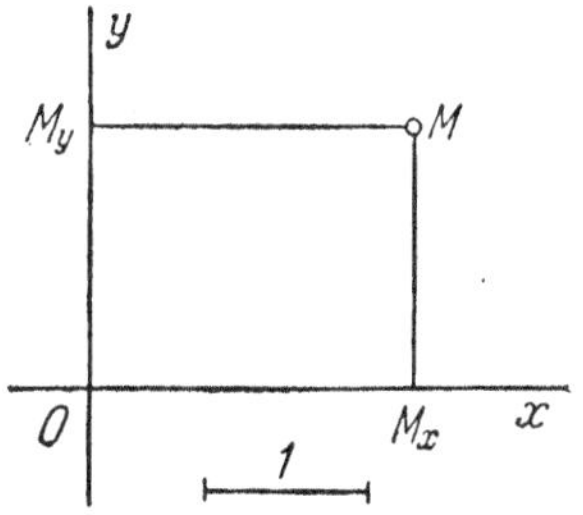

FIG. 1

where M_x, M_y are the projections of the point M on the axes Ox, Oy (Fig. 1) and OM_x, OM_y are the measures of the intervals $\overline{OM_x}$, $\overline{OM_y}$ on the x-axis and y-axis respectively.

The number x may be called the first coordinate of M, and the number y its second coordinate. The symbol $M(x, y)$ means that the point M has the number x as its first coordinate and the number y as its second coordinate.

The y-axis divides the whole plane into two half-planes: the one containing the positive x-axis is called the right half-plane the other the left. Similarly, the x-axis divides the whole plane into two half-planes: the one containing the positive y-axis is called the upper half-plane, the other the lower half-plane.

The two coordinate axes together divide the whole plane into four quadrants, numbered as follows:

1st quadrant lies simultaneously in the right and upper half-planes,
2nd quadrant lies simultaneously in the left and upper half-planes,
3rd quadrant lies simultaneously in the left and lower half-planes,
4th quadrant lies simultaneously in the right and lower half-planes.

17. Construct the points:

$$A(2, 3),\quad B(-5, 1),\quad C(-2, -3),\quad D(0, 3),\quad E(-5, 0),\quad F(-1/3, 2/3).$$

18. Find the coordinates of the projection on to the x-axis of each of the points:

$$A(2, -3),\quad B(3, -1),\quad C(-5, 1),\quad D(-3, -2),\quad E(-5, -1).$$

19. Find the coordinates of the projection on to the y-axis of each of the points:

$$A(-3, 2),\quad B(-5, 1),\quad C(3, -2),\quad D(-1, 1),\quad E(-6, -2).$$

20. Find the coordinates of the points symmetrically situated relative to the x-axis with the points:

$$A(2, 3),\quad B(-3, 2),\quad C(-1, -1),\quad D(-3, -5),\quad E(-4, 6),\quad F(a, b).$$

21. Find the coordinates of the points symmetrically situated relative to the y-axis with the points:

1) $A(-1, 2)$, 2) $B(3, -1)$, 3) $C(-2, -2)$,
4) $D(-2, 5)$, 5) $E(3, -5)$, 6) $F(a, b)$.

22. Find the coordinates of the points symmetrically situated relative to the origin with the points

1) $A(3, 3)$, 2) $B(2, -4)$, 3) $C(-2, 1)$,
4) $D(5, -3)$, 5) $E(-5, -5)$, 6) $F(a, b)$.

23. Find the coordinates of the points which are symmetrical with the given points relative to that bisector of the angle between the coordinate axes which lies in the 1st and 3rd quadrants:

$$1)\ A(2, 3),\quad 2)\ B(5, -2),\quad 3)\ C(-3, 4).$$

24. Find the coordinates of the points which are symmetrical with the given points relative to that bisector of the angle between the coordinate axes which lies in the 2nd and 4th quadrants:

$$1)\ A(3, 5),\quad 2)\ B(-4, 3),\quad 3)\ C(7, -2).$$

25. In which quadrants can the point $M(x, y)$ lie under the following conditions?

1) $xy>0$, 2) $xy<0$, 3) $x-y=0$, 4) $x+y=0$,
5) $x+y>0$ 6) $x+y<0$, 7) $x-y>0$, 8) $x-y<0$.

§ 3. Polar Coordinates

A polar coordinate system is defined when there is given:

(1) a certain point O called the pole,

(2) a ray OA drawn from this point and called the polar axis, and

(3) a scale for measuring lengths. The direction of rotation about O which is to be considered positive must also be specified. The anti-clockwise direction of rotation is generally taken to be positive.

The polar coordinates of an arbitrary point M relative to the given system are the number $\varrho = OM$ and the angle $\theta = \angle AOM$ (Fig. 2). The sign of θ and any added multiple of 2π must be taken into account. The number ϱ is called the first coordinate, the polar radius, or the radius vector; the number θ, measured in radians, is called the second coordinate, the polar angle, or the argument. (The word "amplitude" may also be used.) The symbol $M(\varrho, \theta)$ means that the point M has the polar coordinates ϱ and θ.

(Note that in the preceding paragraph the symbol OM has been used used to denote the length of the interval, taken without regard to sign. There is no need to use the more cumbersome symbol $|OM|$ when, as here, the two points are merely

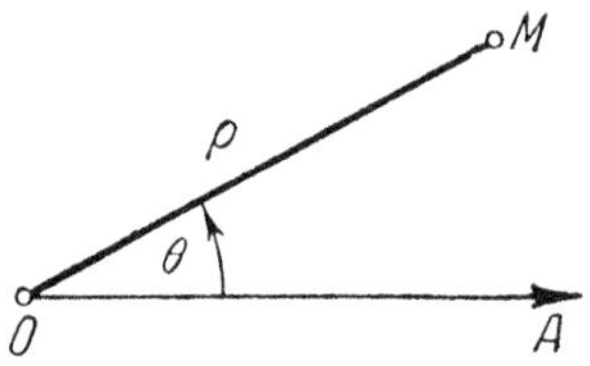

FIG. 2

arbitrary points in the plane and not points on an axis. Similar simplifications in notation will be made in future whenever they do not cause ambiguity.)

Among all the possible values of the polar angle θ (differing from one another by multiples of 2π) the particular one which satisfies the inequality $-\pi < \theta \leqslant +\pi$ is called the principal value.

When Cartesian and polar coordinates are to be used simultaneously, it will be assumed that:

(1) the unit-interval for measurement of lengths is the same for each;

(2) the positive direction of rotation for the polar angle is the same as the direction in which the positive x-axis must be rotated to make it coincide with the positive y-axis by the shortest path. So that, in the usual case with the positive x-axis drawn from left to right and the positive y-axis drawn upwards, the positive sense for the polar angle will be anticlockwise as usual.

If also

(3) the pole of the polar coordinates coincides with the origin of the Cartesian coordinates, and

(4) the polar axis coincides with the positive x-axis, then the Cartesian coordinates can be found from the polar coordinates using the relations:

$$x = \varrho \cos \theta,$$
$$y = \varrho \sin \theta$$

and the polar coordinates can be found from the Cartesians using

$$\varrho = \sqrt{(x^2 + y^2)}, \quad \tan \theta = y/x.$$

Whenever both coordinate systems are used together it will be assumed that the conditions (1) and (2) above are satisfied.

26. Construct the points given by their polar coordinates:

$$A(3, \pi/2), \quad B(2, \pi), \quad C(3, -\pi/4), \quad D(4, 3\tfrac{1}{7}), \quad E(5, 2), \quad F(1, -1).$$

(Construct the points D, E, F approximately using a protractor.)

27. Find the polar coordinates of the points which are symmetrically situated relative to the polar axis with the points given by their polar coordinates:

$$M_1(3, \pi/4), \quad M_2(2, -\pi/2), \quad M_3(3, -\pi/3), \quad M_4(1, 2), \quad M_5(5, -1).$$

28. Find the polar coordinates of the points symmetrical, relative to the pole, with the points:

$$M_1(1, \pi/4), \quad M_2(5, \pi/2), \quad M_3(2, -\pi/3), \quad M_4(4, 5\pi/6), \quad M_5(3, -2).$$

29. In a polar coordinate system the points $A(3, -4\pi/9)$ and $B(5, 3\pi/14)$ are two vertices of the parallelogram $ABCD$ whose diagonals intersect at the pole. Find the polar coordinates of the other two vertices.

30. Find the polar coordinates of the mid-point of the interval $\overline{AB}$, given that

$$A(8, -2\pi/3), \quad B(6, \pi/3).$$

31. In a polar coordinate system the points

$$A(3, \pi/2), \quad B(2, -\pi/4), \quad C(1, \pi), \quad D(5, -3\pi/4), \quad E(3,2), \quad F(2, -1)$$

are given. Find their polar coordinates in a new system obtained by reversing the positive direction of the polar axis.

32. In a polar coordinate system the points

$$M_1(3, \pi/3), \quad M_2(1, 2\pi/3), \quad M_3(2, 0), \quad M_4(5, \pi/4), \quad M_5(3, -2\pi/3), \quad M_6(1, 11\pi/12)$$

are given. The polar axis is rotated to a new position passing through the point M_1. Find the coordinates of the given points in the new polar coordinate system.

33. Find the polar coordinates of the mid-point of the interval M_1M_2, given that $M_1(12, 4\pi/9)$ and $M_2(12, -2\pi/9)$.

34. Calculate the distance between the two points whose polar coordinates are $M_1(\varrho_1, \theta_1)$ and $M_2(\varrho_2, \theta_2)$.

35. Calculate the distance d between the two points whose polar coordinates are $M_1(5, \pi/4)$ and $M_2(8, -\pi/12)$.

36. Calculate the area of a square having as adjacent vertices the points whose polar coordinates are $M_1(12, -\pi/10)$, $M_2(3, \pi/15)$.

37. Calculate the area of a square having as two opposite vertices the points whose polar coordinates are $P(6, -7\pi/12)$, $Q(4, \pi/6)$.

38. Calculate the area of an equilateral triangle having as two vertices the points whose polar coordinates are $A(4, -\pi/12)$ and $B(8, 7\pi/12)$.

39. Calculate the area of the triangle OAB having one vertex at the pole and $A(\varrho_1, \theta_1)$ and $B(\varrho_2, \theta_2)$ as its other vertices.

40. Calculate the area of ΔOAB, given $A(5, \pi/4)$ and $B(4, \pi/12)$.

41. Calculate the area of ΔABC, given the polar coordinates of its vertices: $A(3, \pi/8)$, $B(8, 7\pi/24)$, $C(6, 5\pi/8)$.

42. The pole of a polar coordinate system coincides with the origin of a rectangular Cartesian system, and the polar axis is along the positive x-axis. Find the Cartesian coordinates of the following points, given by their polar coordinates:

$$M_1(6, \pi/2), \quad M_2(5, 0), \quad M_3(2, \pi/4), \quad M_4(10, \pi/3), \quad M_5(8, 2\pi/3), \quad M_6(12, -\pi/6).$$

43. The pole of a polar coordinate system coincides with the origin of a rectangular Cartesian system, and the polar axis is along the positive x-axis. Find the polar coordinates of the following points, given by their Cartesian coordinates:

$$M_1(0, 5), \quad M_2(-3, 0), \quad M_3(\sqrt{3}, 1), \quad M_4(-\sqrt{2}, -\sqrt{2}), \quad M_5(1, \sqrt{3}).$$

§ 4. Directed Intervals. Projection of an Interval on to an Arbitrary Axis. Projection of an Interval on to a Coordinate Axis. Length and Polar Angle of an Interval. Distance between Two Points

A rectilinear interval bounded by two points is said to be directed when it is known which of the two boundary points is to be taken as its beginning and which as its end. A directed interval beginning at point A and ending at point B (see Fig. 3) is denoted by the symbol $\overline{AB}$ (i.e., in the same way as an interval on an axis; see § 1). The length of the directed interval $\overline{AB}$ (for

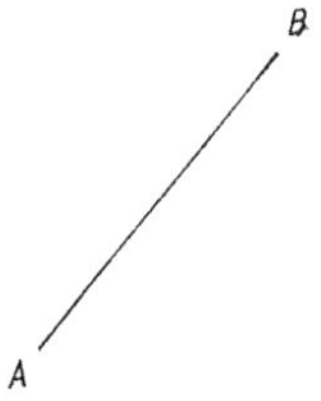

FIG. 3

a given unit-interval) is denoted by the symbol $|AB|$ (or by AB — see note on page 10).

The projection of an interval $\overline{AB}$ on to an axis u is a number equal to the measure of the interval $\overline{A_1B_1}$ on the axis u, where A_1, B_1 are the projections of the points A, B respectively on to the axis u. The projection of an interval $\overline{AB}$ on to the axis u is denoted by the symbol $\text{pr}_u\,\overline{AB}$.

The projection of an interval on to the x-axis of a given rectangular Cartesian coordinate system is denoted by X, and its projection on to the y-axis is denoted by Y.

If the coordinates of the points $M_1(x_1, y_1)$, $M_2(x_2, y_2)$ are known, the projections X, Y of the directed interval $\overline{M_1M_2}$ on to the coordinate axes may be calculated from the equations

$$X = x_2 - x_1, \quad Y = y_2 - y_1.$$

Thus, to find the projection of a directed interval on to either coordinate axis, the coordinate of its beginning is subtracted from the corresponding coordinate of its end.

The angle θ through which the positive x-axis must be turned to give it the same direction as the directed interval $\overline{M_1M_2}$ is called the polar angle of the interval $\overline{M_1M_2}$. The angle θ is to be regarded as in trigonometry, i.e., its sign and any added multiple of 2π is to be taken into account. The value of θ which satisfies the inequality $-\pi < \theta \leqslant +\pi$ is called its principal value.

The equations $X = d\cos\theta$, $Y = d\sin\theta$ give the projections of an arbitrary interval on to the coordinate axes in terms of its length and polar angle. From these relations may be derived the equations expressing the length and polar angle in terms of the projections of the interval on to the coordinate axes, viz.,

$$d = \sqrt{(X^2+Y^2)}, \quad \cos\theta = \frac{X}{\sqrt{(X^2+Y^2)}}, \quad \sin\theta = \frac{Y}{\sqrt{(X^2+Y^2)}}.$$

The distance d between two given points $M_1(x_1, y_1)$, $M_2(x_2, y)$ in a plane is given by

$$d = \sqrt{[(x_2-x_1)^2+(y_2-y_1)^2]}.$$

44. Calculate the projection of an interval on to an axis u, given its length d and angle of inclination φ to the axis:

1) $d = 6$, $\varphi = \pi/3$ 2) $d = 6$, $\varphi = 2\pi/3$ 3) $d = 7$, $\varphi = \pi/2$
4) $d = 5$, $\varphi = 0$ 5) $d = 5$, $\varphi = \pi$ 6) $d = 4$, $\varphi = -\pi/3$.

45. Draw a figure to show the intervals, beginning at the origin, which have the following projections on the coordinate axes:

1) $X = 3$, $Y = 2$ 2) $X = 2$, $Y = -5$
3) $X = -5$, $Y = 0$ 4) $X = 2$, $Y = 3$
5) $X = 0$, $Y = 3$ 6) $X = 5$, $Y = -1$.

46. Draw a figure to show the intervals which begin at the point $M(2, -1)$ and which have the following projections on the coordinate axes:

$$a)\ X=4,\quad Y=3 \qquad b)\ X=2,\ Y=0$$
$$c)\ X=-3,\ Y=1 \qquad d)\ X=4,\ Y=-2$$
$$e)\ X=0,\quad Y=-3 \qquad f)\ X=1,\ Y=-3.$$

47. Given the points $M_1(1, -2)$, $M_2(2, 1)$, $M_3(5, 0)$, $M_4(-1, 4)$, $M_5(0, -3)$, find the projections on to the coordinate axes of the following intervals:

$$1)\ \overline{M_1M_2} \quad 2)\ \overline{M_3M_1} \quad 3)\ \overline{M_4M_5} \quad 4)\ \overline{M_5M_3}.$$

48. The projections of the interval $\overline{M_1M_2}$ on the coordinate axes are $X=5$, $Y=-4$, respectively, and the interval begins at the point $M_1(-2, 3)$. Find the coordinates of its end.

49. The projections on to the coordinate axes of the interval $\overline{AB}$ are $X=4$, $Y=-5$, and the interval ends at the point $B(1, -3)$. Find the coordinates of its beginning.

50. Draw a figure to show the intervals which start at the origin and have length d and polar angle θ if

$$1)\ d=5,\ \theta=+\pi/5 \quad 2)\ d=3,\ \theta=5\pi/6$$
$$3)\ d=4,\ \theta=-\pi/3 \quad 4)\ d=3,\ \theta=-4\pi/3.$$

51. Draw a figure to show the intervals which start at the point $M(2, 3)$ and have length d and polar angle θ if

$$1)\ d=2,\ \theta=-\pi/10 \quad 2)\ d=1,\ \theta=\pi/9$$
$$3)\ d=5,\ \theta=-\pi/2$$

(M is given by its Cartesian coordinates.)

52. Calculate the projections on to the coordinate axes of the intervals whose length d and polar angle θ are

1) $d = 12$, $\theta = 2\pi/3$, 2) $d = 6$, $\theta = -\pi/6$,
3) $d = 2$, $\theta = -\pi/4$.

53. Calculate the lengths of the intervals whose projections on to the coordinate axes are

1) $X = 3$, $Y = -4$, 2) $X = 12$, $Y = 5$,
3) $X = -8$, $Y = 6$.

54. Calculate the length d and polar angle θ of each of the intervals whose projections on to the coordinate axes are

1) $X = 1$, $Y = \sqrt{3}$, 2) $X = 3\sqrt{2}$, $Y = -3\sqrt{2}$,
3) $X = -2\sqrt{3}$, $Y = 2$.

55. Given the points $M_1(2, -3)$, $M_2(1, -4)$, $M_3(-1, -7)$, $M_4(-4, 8)$, calculate the length and polar angle of the following intervals:

a) $\overline{M_1M_2}$, b) $\overline{M_1M_3}$, c) $\overline{M_2M_4}$, d) $\overline{M_4M_3}$.

56. The length of an interval is $d = 5$ and its projection on the x-axis equals 4. Find its projection on the y-axis if it makes with the y-axis a) an acute angle, b) an obtuse angle.

57. The length of the interval $\overline{MN}$ equals 13, it begins at the point $M(3, -2)$, and its projection on the x-axis equals -12. Find the coordinates of the end of this interval if it makes with the y-axis a) an acute angle, b) an obtuse angle.

58. The length of the interval $\overline{MN}$ equals 17, it ends at the point $N(-7, 3)$, and its projection on the y-axis equals 15. Find the coordinates of the beginning of the interval if it makes with the x-axis a) an acute angle, b) an obtuse angle.

59. If the projections of an interval on the coordinate axes are $X = 1$, $Y = -\sqrt{3}$, find its projection on an axis which makes with the x-axis an angle $\theta = 2\pi/3$.

60. Given the two points $M_1(1, -5)$, $M_2(4, -1)$, find the projection of the interval M_1M_2 on an axis which makes with the x-axis an angle $\theta = -\pi/6$.

61. Given the two points $P(-5, 2)$, $Q(3, 1)$, find the projection of the interval $\overline{PQ}$ on an axis which makes with the x-axis an angle $\theta = \tan^{-1} 4/3$.

62. Given the two points $M_1(2, -2)$, $M_2(7, -3)$, find the projection of the interval $\overline{M_1M_2}$ on an axis passing through the points $A(5, -4)$, $B(-7, 1)$ in the direction 1) from A to B, 2) from B to A.

63. Given the points $A(0, 0)$, $B(3, -4)$, $C(-3, 4)$, $D(-2, 2)$, $E(10, -3)$, find the distance d between the points: 1) A, B 2) B, C 3) A, C 4) C, D 5) A, D 6) D, E.

64. Calculate the area of a square having $A(3, -7)$ and $B(-1, 4)$ as adjacent vertices.

65. Calculate the area of the square having $P(3, 5)$ and $Q(1, -3)$ as opposite vertices.

66. Calculate the area of an equilateral triangle having $A(-3, 2)$ and $B(1, 6)$ as two vertices.

67. Calculate the lengths of the diagonals of the parallelogram $ABCD$ if three of its vertices are the points $A(3, -7)$, $B(5, -7)$, $C(-2, 5)$ and D and B are opposite vertices.

68. Calculate the area of the rhombus having a side equal in length to $5\sqrt{10}$ and having $P(4, 9)$, $Q(-2, 1)$ as opposite vertices.

69. Calculate the length of an altitude of the rhombus which has a side equal in length to $5\sqrt{2}$ and the points $P(3, -4)$, $Q(1, 2)$ as opposite vertices.

70. Prove that the points $A(3, -5)$, $B(-2, -7)$, $C(18, 1)$ lie on a straight line.

71. Prove that the triangle having the points $A_1(1, 1)$, $A_2(2, 3)$, $A_3(5, -1)$ as vertices is right-angled.

72. Prove that the points $A(2, 2)$, $B(-1, 6)$, $C(-5, 3)$, $D(-2, -1)$ are the vertices of a square.

73. Is the triangle which has as vertices the points $M_1(1, 1)$, $M_2(0, 2)$, $M_3(2, -1)$ an obtuse-angled triangle?

74. Prove that the internal angles of the triangle with vertices at $M(-1, 3)$, $N(1, 2)$, $P(0, 4)$ are all acute.

75. Calculate the internal angles of a triangle having as vertices the points $A(5, 0)$, $B(0, 1)$, $C(3, 3)$.

76. The vertices of a triangle are the points $A(-\sqrt{3}, 1)$, $B(0, 2)$, $C(-2\sqrt{3}, 2)$: calculate its exterior angle at the vertex A.

77. Find a point M on the x-axis whose distance from the point $N(2, -3)$ is equal to 5.

78. Find a point M on the y-axis whose distance from the point $N(-8, 13)$ is equal to 17.

79. Given the two points $M(2, 2)$, $N(5, -2)$, find a point P on the x-axis such that $\angle MPN$ is a right-angle.

80. Find the centre C and radius R of the circle drawn through the point $A(4, 2)$ to touch both coordinate axes.

81. Find the centre C of the circle with radius 5 drawn through the point $M_1(1, -2)$ to touch the x-axis.

82. Find the coordinates of the point M_2 which is the mirror-image of the point $M_1(1, 2)$ in the straight line passing through the points $A(1, 0)$ and $B(-1, -2)$.

83. $A(3, 0)$ and $C(-4, 1)$ are two opposite vertices of a square: find its other two vertices.

84. $A(2, -1)$ and $B(-1, 3)$ are two adjacent vertices of a square: find its other two vertices.

85. $M_1(-3, 6)$, $M_2(9, -10)$, $M_3(-5, 4)$ are the vertices of a triangle: find the centre C and the radius R of the circum-circle of the triangle.

§ 5. Division of an Interval in a Given Ratio

If the point $M(x, y)$ lies on the straight line passing through two given points $M_1(x_1, y_1)$ and $M_2(x_2, y_2)$, and if the ratio $\lambda = M_1M/MM_2$ in which the point M divides the interval $\overline{M_1M_2}$ is given, then the coordinates of the point M are defined by the relations

$$x = \frac{x_1 + \lambda x_2}{1+\lambda}, \quad y = \frac{y_1 + \lambda y_2}{1+\lambda}.$$

If M is the mid-point of the interval $\overline{M_1M_2}$, then its coordinates are given by

$$x = (x_1 + x_2)/2, \quad y = (y_1 + y_2)/2.$$

86. $A(3, -5)$, $B(-1, 1)$ are the ends of a uniform beam: find its centre of gravity.

87. The centre of gravity of a uniform beam is at the point $M(1, 4)$ and one of its ends is at the point $P(-2, 2)$. Find the coordinates of its other end Q.

88. $A(1, -3)$, $B(3, -5)$, $C(-5, 7)$ are the vertices of a triangle. Find the coordinates of the mid-points of its sides.

89. Given the points $A(3, -1)$, $B(2, 1)$, find:

(1) the coordinates of the point M symmetrical with A relative to B,

(2) the coordinates of the point N symmetrical with B relative to A.

90. The points $M(2, -1)$, $N(-1, 4)$, $P(-2, 2)$ are the mid-points of the sides of a certain triangle: find its vertices.

91. $A(3, -5)$, $B(5, -3)$, $C(-1, 3)$ are three vertices of a parallelogram. Find its fourth vertex D, opposite to B.

92. $A(-3, 5)$, $B(1, 7)$ are two adjacent vertices of a parallelogram, and $M(1, 1)$ is the point of intersection of its diagonals. Find its other two vertices.

93. $A(2, 3)$, $B(4, -1)$, $C(0, 5)$ are three vertices of the parallelogram $ABCD$: find its fourth vertex D.

94. $A(1, 4)$, $B(3, -9)$, $C(-5, 2)$ are the vertices of a triangle: find the length of its median drawn through the point B.

95. Find the coordinates of the points of trisection of the interval bounded by the points $A(1, -3)$, $B(4, 3)$.

96. $A(2, -5)$, $B(1, -2)$, $C(4, 7)$ are the vertices of a triangle; find the point of intersection of the internal bisector of the angle B with the side AC.

97. $A(3, -5)$, $B(-3, 3)$, $C(-1, -2)$ are the vertices of a triangle: find the length of the internal bisector of the angle A.

98. $A(-1, -1)$, $B(3, 5)$, $C(-4, 1)$ are the vertices of a triangle: find the point of intersection of the external bisector of the angle A with the extension of the side BC.

99. $A(3, -5)$, $B(1, -3)$, $C(2, -2)$ are the vertices of a triangle: find the length of the external bisector of the angle B.

100. The three points $A(1, -1)$, $B(3, 3)$, $C(4, 5)$ lie on a straight line. Find the ratio λ in which each of them divides the interval bounded by the other two.

101. Find the coordinates of the ends A and B of the interval whose points of trisection are the points $P(2, 2)$ and $Q(1, 5)$.

102. A straight line passes through the points $M_1(-12, -13)$, $M_2(-2, -5)$. Find the point on this line whose x-coordinate is 3.

103. A straight line passes through the points $M(2, -3)$, $N(-6, 5)$. Find the point on this line whose y-coordinate is -5.

104. Find the point of intersection with the x-axis of the straight line passing through the points $A(7, -3)$, $B(23, -6)$.

105. Find the point of intersection with the y-axis of the straight line passing through the points $A(5, 2)$, $B(-4, -7)$.

106. $A(-3, 12)$, $B(3, -4)$, $C(5, -4)$, $D(5, 8)$ are the vertices of a quadrangle. Find the ratio in which the diagonal AC divides the diagonal BD.

107. $A(-2, 14)$, $B(4, -2)$, $C(6, -2)$, $D(6, 10)$ are the vertices of a quadrangle. Find the point of intersection of its diagonals AC and BD.

108. $A(x_1, y_1)$, $B(x_2, y_2)$, $C(x_3, y_3)$ are the vertices of a uniform triangular lamina. Find the coordinates of its centre of gravity.

Note. The centre of gravity coincides with the point of intersection of the medians.

109. The point of intersection M of the medians of a triangle lies on the x-axis. $A(2, -3)$, $B(-5, 1)$ are two vertices of the triangle, and its third vertex C lies on the y-axis. Find the coordinates of the points M and C.

110. $A(x_1, y_1)$, $B(x_2, y_2)$, $C(x_3, y_3)$ are the vertices of a uniform triangular lamina. By joining the mid-points of its sides, a new uniform triangular lamina is obtained. Prove that the centres of gravity of the two laminae coincide.

Hint. Use the result of question 108.

111. A uniform lamina has the form of a square, of side 12, from which one-quarter has been cut away along the lines joining the mid-points of two of its sides to its centre of gravity, as shown in Fig. 4. The axes are along the sides of the square. Find the centre of gravity of the lamina.

112. A uniform lamina has the form of a rectangle, of sides a and b, from which one-quarter has been cut away along the lines joining the mid-points of two of its sides to its centre of gravity, as shown in Fig. 5. The axes are along the sides of the rectangle. Find the centre of gravity of the lamina.

113. A uniform lamina has the form of a square, of side $2a$, from which a triangle has been cut away along a line joining

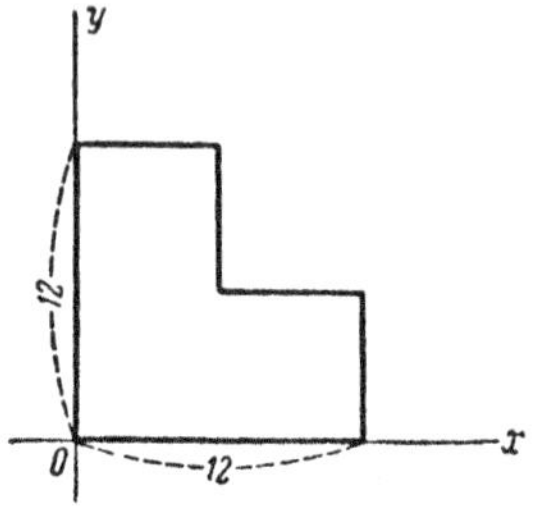

FIG. 4

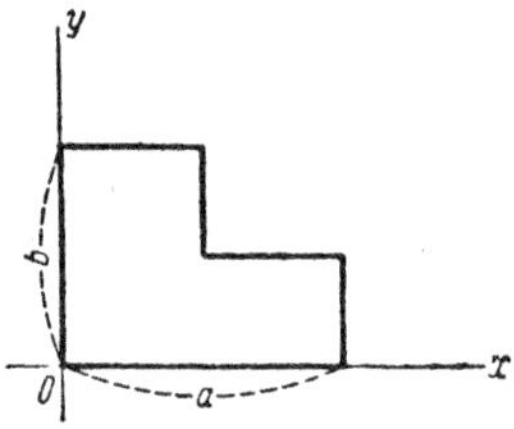

FIG. 5

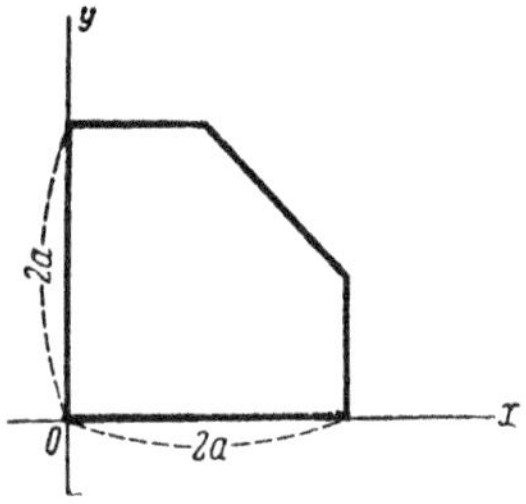

FIG. 6

the mid-points of two adjacent sides, as shown in Fig. 6. The axes are along the sides of the square. Find the centre of gravity of the lamina.

114. Find the centre of gravity of masses m, n, and p concentrated at the points $A(x_1, y_1)$, $B(x_2, y_2)$, $C(x_3, y_3)$ respectively.

115. $A(4, 2)$, $B(7, -2)$, $C(1, 6)$ are the vertices of a triangle constructed from uniform wire. Find the centre of gravity.

§ 6. Area of a Triangle

The area S of the triangle ABC formed by any three points $A(x_1, y_1)$, $B(x_2, y_2)$, $C(x_3, y_3)$ is given by

$$\pm S = \frac{1}{2}\begin{vmatrix} x_2 - x_1 & y_2 - y_1 \\ x_3 - x_1 & y_3 - y_1 \end{vmatrix}$$

the sign being plus when the shortest rotation from the interval $\overline{AB}$ to the interval $\overline{AC}$ is positive, and minus when this rotation is negative.

116. Calculate the area of the triangles whose vertices are:

1) $A(2, -3)$, $B(3, 2)$, $C(-2, 5)$
2) $M_1(-3, 2)$, $M_2(5, -2)$, $M_3(1, 3)$
3) $M(3, -4)$, $N(-2, 3)$, $P(4, 5)$.

117. $A(3, 6)$, $B(-1, 3)$, $C(2, -1)$ are the vertices of a triangle. Find the length of its altitude drawn through the vertex C.

118. Find the area of the parallelogram three of whose vertices are the points $A(-2, 3)$, $B(4, -5)$, $C(-3, 1)$.

119. $A(3, 7)$, $B(2, -3)$, $C(-1, 4)$ are three vertices of a parallelogram. Find the length of its altitude drawn from the vertex B on to the side AC.

120. Find the centre of gravity of a uniform quadrangular lamina whose vertices are the points $A(2, 1)$, $B(5, 3)$, $C(-1, 7)$, $D(-7, 5)$.

121. Find the centre of gravity of a uniform lamina in the shape of a pentagon with vertices at the points $A(2, 3)$, $B(0, 6)$, $C(-1, 5)$, $D(0, 1)$ and $E(1, 1)$.

122. The area of a triangle is $S = 3$. Two of its vertices are the points $A(3, 1)$, $B(1, -3)$, and its third vertex C lies on the y-axis. Find the coordinates of the vertex C.

123. The area of a triangle is $S = 4$. Two of its vertices are the points $A(2, 1)$, $B(3, -2)$, and its third vertex C lies on the x-axis. Find the coordinates of the vertex C.

124. The area of a triangle is $S = 3$. Two of its vertices are the points $A(3, 1)$, $B(1, -3)$, and its centre of gravity lies on the x-axis. Find the coordinates of its third vertex C.

125. The area of a parallelogram is $S = 12$. Two of its vertices are the points $A(-1, 3)$, $B(-2, 4)$, and its diagonals intersect on the x-axis. Find its other two vertices.

126. The area of a parallelogram is $S = 17$. Two of its vertices are the points $A(2, 1)$, $B(5, -3)$ and its diagonals intersect on the y-axis. Find its other two vertices.

§ 7. Transformation of Coordinates

The formulae for transformation of rectangular Cartesian coordinates when the position of the origin is changed but the directions of the axes and also the unit-interval remain unchanged are

$$x = \tilde{x} + a, \quad y = \tilde{y} + b,$$

where x, y are the coordinates of an arbitrary point M of the plane relative to the old axes, $\tilde{x}$, $\tilde{y}$ are its coordinates relative to the new axes, and a, b are the coordinates of the new origin $\tilde{O}$ relative to the old axes (a is also the measure of the displacement of the origin in the direction of the axis Ox, and b is the measure of the displacement of the origin in the direction of the axis Oy).

The formulae for transformation of rectangular Cartesian coordinates when the axes are rotated through an angle α (which must be understood as in trigonometry) are

$$x = x' \cos\alpha - y' \sin\alpha,$$
$$y = x' \sin\alpha + y' \cos\alpha,$$

where x, y are the coordinates of an arbitrary point M of the plane relative to the old axes, and x', y' are its coordinates relative to the new axes.

The formulae

$$x = \tilde{x}' \cos\alpha - \tilde{y}' \sin\alpha + a,$$
$$y = \tilde{x}' \sin\alpha + \tilde{y}' \cos\alpha + b$$

give the transformation of coordinates when the axes are first moved to a parallel position through a displacement of measure a in the direction Ox and of measure b in the direction Oy and are then rotated through the angle α.

The unit-interval or scale is assumed to remain unchanged in each of the above transformations, and also in the following examples.

127. Write down the formulae for transformation of coordinates if the origin is transferred, without change of direction of the axes, to the points: 1) $A(3, 4)$, 2) $B(-2, 1)$, 3) $C(-3, 5)$.

128. The origin is transferred to the point $\tilde{O}(3, -4)$ without changing the directions of the axes. The points A, B, C have the coordinates $A(1, 3)$, $B(-3, 0)$, $C(-1, 4)$ relative to the new axes. Find their old coordinates.

129. The three points $A(2, 1)$, $B(-1, 3)$, $C(-2, 5)$ are given. Find their new coordinates if the origin is transferred (without change of direction of the axes) to: 1) the point A, 2) the point B, 3) the point C.

130. Find the old coordinates of the origin $\tilde{O}$ of the new system if the transformation formulae are:

1) $x = \tilde{x}+3,\ y = \tilde{y}+5$ 2) $x = \tilde{x}-2,\ y = \tilde{y}+1$
3) $x = \tilde{x},\ y = \tilde{y}-1$ 4) $x = \tilde{x}-5,\ y = \tilde{y}$.

131. Write down the formulae for transformation of coordinates if the axes are rotated through the angle:

1) 60° 2) −45° 3) 90° 4) −90° 5) 180°.

132. The coordinate axes are rotated through the angle $\alpha = 60°$. The points A, B, C in the new system have the coordinates $A(2\sqrt{3}, -4)$, $B(\sqrt{3}, 0)$, $C(0, -2\sqrt{3})$. Find their coordinates in the old system.

133. The points $M(3, 1)$, $N(-1, 5)$, $P(-3, -1)$ are given. Find their new coordinates if the axes are rotated through the angle:

1) −45° 2) 90° 3) −90° 4) 180°

134. Find the angle α through which the axes have been rotated if the transformation formulae are:

1) $x = x'/2 - y'\sqrt{3}/2,\ y = x'\sqrt{3}/2 + y'/2$
2) $x = x'\sqrt{3}/2 + y'/2,\ y = -x'/2 + y'\sqrt{3}/2$.

135. Find the coordinates of $\tilde{O}$, the new coordinate origin, if the point $A(3, -4)$ lies on the new x-axis and the point $B(2,3)$ lies on the new y-axis, and corresponding axes of the old and new systems are parallel.

136. Write down the formulae for transformation of coordinates if the point $M_1(2, -3)$ lies on the new x-axis and the point $M_2(1, -7)$ lies on the new y-axis, and corresponding axes of the old and new systems are parallel.

137. Two systems of coordinate axes Ox, Oy and Ox', Oy' have a common origin O and transform into one another by rotation through a certain angle. The coordinates of the point $A(3, -4)$ are defined relative to first system. Deduce the transformation formulae, given that the positive direction of the axis Ox' is defined by the interval $\overline{OA}$.

138. The coordinate origin is displaced to the point $\tilde{O}(-1, 2)$, and the coordinate axes are rotated through the angle $\alpha = \tan^{-1} 5/12$. The coordinates of the points $M_1(3, 2)$, $M_2(2, -3)$, $M_3(13, -13)$ refer to the new system. Find the coordinates of these points referred to the old system.

139. $A(5, 5)$, $B(2, -1)$, $C(12, -6)$ are three given points. Find their coordinates in a new system obtained by displacing the origin to the point B and rotating the axes through the angle $\alpha = \tan^{-1} 3/4$.

140. Find the old coordinates of the new origin and the angle α through which the axes have been rotated if the transformation formulae are:

1) $x = -\tilde{y}' + 3$, $y = \tilde{x}' - 2$ 2) $x = -\tilde{x}' - 1$, $y = -\tilde{y}' + 3$
3) $x = \tilde{x}' \sqrt{2}/2 + \tilde{y}' \sqrt{2}/2 + 5$, $y = -\tilde{x}' \sqrt{2}/2 + \tilde{y}' \sqrt{2}/2 - 3$.

141. $M_1(9, -3)$, $M_2(-6, 5)$ are two given points. Find the transformation formulae if the origin is displaced to the point M_1 and the axes are rotated so that the positive x-axis is along the directed interval $\overline{M_1M_2}$.

142. The polar axis of a polar coordinate system is parallel to the x-axis of a rectangular Cartesian coordinate system and has the same direction. The Cartesian coordinates of the pole are $O(1, 2)$. Find the Cartesian coordinates of the points whose polar coordinates are

$$M_1(7, \pi/2),\ M_2(3, 0),\ M_3(5, -\pi/2),\ M_4(2, 2\pi/3),\ M_5(2, -\pi/6).$$

143. The pole of a polar coordinate system coincides with the origin of a rectangular Cartesian coordinate system, and the polar axis lies in the first quadrant and makes equal angles with the x-axis and y-axis. Find the Cartesian coordinates of the points whose polar coordinates are

$$M_1(5, \pi/4),\ M_2(3, -\pi/4),\ M_3(1, 3\pi/4),\ M_4(6, -3\pi/4),\ M_5(2, -\pi/12).$$

144. The polar axis of a polar coordinate system is parallel to, and has the same direction as, the x-axis of a rectangular Cartesian coordinate system. In Cartesian coordinates, the pole is at $O(3, 2)$ and $M_1(5, 2)$, $M_2(3, 1)$, $M_3(3, 5)$, $M_4(3+\sqrt{2}, 2-\sqrt{2})$, $M_5(3+\sqrt{3}, 3)$ are given points. Find the polar coordinates of these points.

145. The pole of a polar coordinate system coincides with the origin of a rectangular Cartesian coordinate system, and the polar axis lies in the first quadrant and makes equal angles with the x-axis and y-axis. Find the polar coordinates of the points whose Cartesian coordinates are

$$M_1(-1, 1),\ M_2(\sqrt{2}, -\sqrt{2}),\ M_3(1, \sqrt{3}),\ M_4(-\sqrt{3}, 1),\ M_5(2\sqrt{3}, -2).$$

CHAPTER 2

EQUATION OF A CURVE

§ 8. Functions of Two Variables

If some rule has been given which assigns to each point M of a plane (or of some part of the plane) a certain number u, then we say that in the plane (or part of the plane) a "function of position" has been given. The fact that u is a function of the position M is expressed symbolically by an equality of the form $u = f(M)$. The number u prescribed at the point M is said to be the value of the given function at the point M. For example, if A is a fixed point, and M an arbitrary point, in the plane, then the distance from A to M is a function of the position M, and in this case $f(M) = AM$.

Suppose some function $u = f(M)$ is given, and that a system of coordinates has been introduced, so that an arbitrary point M is defined by its coordinates x, y. Then the value of the given function is also determined by the coordinates x, y; alternatively we say that $u = f(M)$ is a function of the two variables x, y. A function of these two variables is denoted by the symbol $f(x, y)$. If $f(M) = f(x, y)$, then the function $u = f(x, y)$ is called a representation of the given function in the chosen coordinate system.

Thus, in the previous example, $f(M) = AM$, and if we introduce a rectangular Cartesian coordinate system with origin at the point A, then we obtain as a representation of this function $u = \sqrt{(x^2 + y^2)}$.

146. Given two points P and Q, separated by a distance a, and a function $f(M) = d_1^2 - d_2^2$, where $d_1 = MP$ and $d_2 = MQ$, find an expression for the representation of this function in a coordinate system having its origin at the point P and the x-axis directed along the interval $\overline{PQ}$.

147. Obtain expressions for the representation of the function defined in question **146** (both directly and also using the transformation formulae and the result of q. **146**) in the coordinate systems:

(1) having the origin at the mid-point of the interval $\overline{PQ}$, and the x-axis directed along the interval $\overline{PQ}$,

(2) having the origin at P, and the x-axis directed along the interval $\overline{PQ}$.

148. Given a square $ABCD$ of side a and a function

$$f(M) = d_1^2 + d_2^2 + d_3^2 + d_4^2,$$

where $d_1 = MA, \quad d_2 = MB, \quad d_3 = MC, \quad d_4 = MD,$

find an expression for the representation of this function in a coordinate system having its axes along the diagonals of the square (Ox being directed along the interval $\overline{AC}$ and Oy along the interval $\overline{BD}$.)

149. Obtain an expression for the representation of the function defined in q. **148** (both directly and also using the transformation formulae and the result of q. **148**) in a coordinate system having its origin at the point A and the axes along the sides of the square (the x-axis along the interval $\overline{AB}$, and the y-axis along the interval $\overline{AD}$).

150. Given the function $f(x, y) = x^2 + y^2 - 6x + 8y$, find an expression for it in a new coordinate system obtained by shifting the origin to the point $\tilde{O}(3, -4)$ without changing the direction of the axes.

151. Given the function $f(x, y) = x^2 - y^2 - 16$, find an expression for it in a new coordinate system obtained by rotating the axes through an angle $-45°$.

152. Given the function $f(x, y) = x^2 + y^2$, find an expression for it in a new coordinate system obtained by rotating the axes through an angle α.

153. Find a point such that when the origin is transferred to it, the transformed expression for the function $f(x, y) = x^2 - 4y^2 - 6x + 8y + 3$ in the new coordinate system shall contain no first-degree terms.

154. Find a point such that when the origin is transferred to it, the transformed expression for the function

$$f(x, y) = x^2 - 4xy + 4y^2 + 2x + y - 7$$

in the new coordinate system shall contain no first-degree terms.

155. Through what angle must the coordinate axes be rotated so that the expression for the function

$$f(x, y) = x^2 - 2xy + y^2 - 6x + 3$$

after transformation shall contain no cross product term in the new variables?

156. Through what angle must the coordinate axes be rotated so that the expression for the function

$$f(x, y) = 3x^2 + 2\sqrt{3}xy + y^2$$

after transformation shall contain no product term in the new variables?

§ 9. The Concept of an Equation of a Curve. The Specification of a Curve by Means of an Equation

An equality of the form $F(x, y) = 0$ is called an equation in the two variables x and y, provided only that it does not hold for every pair of numbers x and y. Two numbers $x = x_0$ and $y = y_0$ are said to satisfy an equation of the form $F(x, y) = 0$ if the left-hand member becomes equal to zero when the variables x, y are replaced by the numbers x_0, y_0.

An equation of a given curve (in a chosen system of coordinates) is an equation in two variables which is satisfied by the coordinates of every point which lies on the curve and which is not satisfied by the coordinates of any point which does not lie on the curve.

If $F_1(x, y) = 0$ and $F_2(x, y) = 0$ are the equations of two curves, then the simultaneous solution of the equations

$$F_1(x, y) = 0, \quad F_2(x, y) = 0$$

gives all the points of intersection of the two curves. More precisely, any pair of numbers which satisfy the two equations simultaneously defines one of the points of intersection of the curves.

NOTE: The coordinate system is to be assumed to be a rectangular Cartesian system in all cases unless some other system is specified.

157. Which of the points $M_1(2, -2)$, $M_2(2, 2)$, $M_3(2, -1)$, $M_4(3, -3)$, $M_5(5, -5)$, $M_6(3, -2)$ lie on the curve defined by the equation $x + y = 0$? What sort of curve is defined by this equation? Sketch it.

158. On the curve defined by the equation $x^2 + y^2 = 25$ find the points whose x-coordinates are equal to: a) 0, b) -3, c) 5, d) 7, and the points whose y-coordinates are equal to:

e) 3, f) -5, g) -8. What sort of curve is defined by this equation? Sketch it.

159. What sort of curve is defined by each of the following equations? Construct the curves on graph-paper.

1) $x-y=0$.
2) $x+y=0$.
3) $x-2=0$.
4) $x+3=0$.
5) $y-5=0$.
6) $y+2=0$.
7) $x=0$.
8) $y=0$.
9) $x^2-xy=0$.
10) $xy+y^2=0$.
11) $x^2-y^2=0$.
12) $xy=0$.
13) $y^2-9=0$.
14) $x^2-8x+15=0$.
15) $y^2+5y+4=0$.
16) $x^2y-7xy+10y=0$.
17) $y=|x|$.
18) $x=|y|$.
19) $y+|x|=0$.
20) $x+|y|=0$.
21) $y=|x-1|$.
22) $y=|x+2|$.
23) $x^2+y^2=16$.
24) $(x-2)^2+(y-1)^2=16$.
25) $(x+5)^2+(y-1)^2=9$.
26) $(x-1)^2+y^2=4$.
27) $x^2+(y+3)^2=1$.
28) $(x-3)^2+y^2=0$.
29) $x^2+2y^2=0$.
30) $2x^2+3y^2+5=0$.
31) $(x-2)^2+(y+3)^2+1=0$.

160. Which of the following curves goes through the origin?

1) $x+y=0$. 2) $x-y=0$. 3) $x^2+y^2-36=0$.
4) $x^2+y^2-2x+y=0$. 5) $x^2+y^2+4x-6y-1=0$.

161. Find the points of intersection of the following curves (a) with the x-axis, (b) with the y-axis:

1) $x^2+y^2=49$.
2) $(x-3)^2+(y+4)^2=25$.
3) $(x+6)^2+(y-3)^2=25$.
4) $(x+5)^2+(y-4)^2=9$.
5) $x^2+y^2-12x+16y=0$.
6) $x^2+y^2-2x+8y+7=0$.
7) $x^2+y^2-6x+4y+12=0$.

162. Find the points of intersection of the two curves given by:

1) $x^2+y^2 = 8,\ x-y = 0.$

2) $x^2+y^2-16x+4y+18 = 0,\ x+y = 0.$

3) $x^2+y^2-2x+4y-3 = 0,\ x^2+y^2 = 25.$

4) $x^2+y^2-8x+10y+40 = 0,\ x^2+y^2 = 4.$

163. In a polar coordinate system, which of the points $M_1(1, \pi/3)$, $M_2(2, 0)$, $M_3(2, \pi/4)$, $M_4(\sqrt{3}, \pi/6)$, $M_5(1, 2\pi/3)$ lie on the curve given by $\varrho = 2\cos\theta$? What sort of curve is defined by this equation? Draw its graph.

164. A curve is defined in a polar coordinate system by the equation $\varrho = 3/\cos\theta$. Find the points of the curve whose polar angles are equal to: a) $\pi/3$, b) $-\pi/3$, c) 0, d) $\pi/6$. What sort of curve is defined by this equation? Draw its graph.

165. A curve is defined in a polar coordinate system by the equation $\varrho = 1/\sin\theta$. Find the points of the curve whose polar radii are equal to: a) 1, b) 2, c) $\sqrt{2}$. What sort of curve is defined by this equation? Draw its graph.

166. What sort of curve is defined in a polar coordinate system by each of the following equations? Draw the graph of each.

1) $\varrho = 5$, 2) $\theta = \pi/3$, 3) $\theta = -\pi/4$, 4) $\varrho\cos\theta = 2$,

5) $\varrho\sin\theta = 1$, 6) $\varrho = 6\cos\theta$, 7) $\varrho = 10\sin\theta$,

8) $\sin\theta = 1/2$, 9) $\sin\varrho = 1/2$.

167. Draw the graphs of the Archimedean spirals given in polar coordinates by the equations: 1) $\varrho = 2\theta$, 2) $\varrho = 5\theta$, 3) $\varrho = \theta/\pi$, 4) $\varrho = -\theta/\pi$.

168. Draw the graphs of the hyperbolic spirals given in polar coordinates by the equations: 1) $\varrho = 1/\theta$, 2) $\varrho = 5/\theta$, 3) $\varrho = \pi/\theta$, 4) $\varrho = -\pi/\theta$.

169. Draw the graphs of the logarithmic spirals given in polar coordinates by the equations: 1) $\varrho = 2^\theta$, 2) $\varrho = (1/2)^\theta$.

170. Find the length of the intervals in which the Archimedean spiral given by $\varrho = 3\theta$ cuts the ray starting at the origin which is inclined at an angle $\theta = \pi/6$ to the polar axis. Draw the figure.

171. The point C whose polar radius is equal to 47 is taken on the Archimedean spiral given by $\varrho = 5\theta/\pi$. Into how many parts does the spiral divide the radius OC? Draw the figure.

172. Find the point P whose polar radius is equal to 12 on the hyperbolic spiral given by $\varrho = 6/\theta$. Draw the figure.

173. Find the point Q whose polar radius is equal to 81 on the logarithmic spiral given by $\varrho = 3^\theta$. Draw the figure.

§ 10. Derivation of Equations for Previously Defined Curves

In the last section a curve was defined by giving its equation. Here we have the opposite sort of problem: a curve is defined purely geometrically and we have to find its equation.

EXAMPLE 1. Derive equation in a rectangular Cartesian coordinate system for the locus of a point so that the sum of the squares of its distances from two fixed points $A_1(-a, 0)$ and $A_2(a, 0)$ is always equal to $4a^2$.

Solution. Let $M(x, y)$ be an arbitrary point of the curve; since M can be anywhere on the curve, the x and y are variables, called the current coordinates.

The geometrical property of the curve can be expressed symbolically by

$$(MA_1)^2+(MA_2)^2 = 4a^2. \tag{1}$$

As M moves the lengths MA_1 and MA_2 in this relation vary. We express them in terms of the current coordinates of the point M:

$$MA_1 = \sqrt{[(x+a)^2+y^2]}, \quad MA_2 = \sqrt{[(x-a)^2+y^2]}.$$

Substituting these expressions in the relation (1), we find the equation between the coordinates x, y of the point M:

$$(x+a)^2+y^2+(x-a)^2+y^2 = 4a^2. \tag{2}$$

This is the required equation of the given curve.

For, any point M which lies on this curve will satisfy the condition (1) and consequently the coordinates of M will satisfy the equation (2). But any point M which does not lie on the curve will not satisfy condition (1) and consequently its coordinates will not satisfy the equation (2).

The problem is therefore solved, but we notice that the equation (2) can be simplified. Removing the brackets and collecting similar terms, we get the equation of the given curve in the form

$$x^2+y^2 = a^2$$

and it is now easy to see that the given curve is a circle with its centre at the origin and radius equal to a.

EXAMPLE 2. Derive the equation in a polar coordinate system for the circle with centre $C(\varrho_0, \theta_0)$ and radius r (Fig. 7).

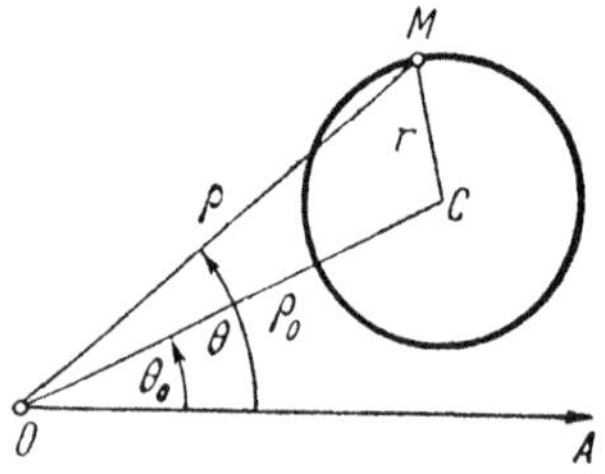

FIG. 7

Solution. Let $M(\varrho, \theta)$ be an arbitrary point of the circle; since M can be anywhere on the circle, the ϱ and θ are variables, the current coordinates.

All points of the circle are at the distance r from its centre; this condition may be written as

$$CM = r. \tag{1}$$

We express CM in terms of the current coordinates, by using the cosine-formula for a triangle (see Fig. 7):

$$CM = \sqrt{[\varrho^2 + \varrho_0^2 - 2\varrho_0\varrho \cos(\theta - \theta_0)]}.$$

Putting this expression in the equality (1), we find the equation between the coordinates ϱ, θ of the point M:

$$\sqrt{[\varrho^2 + \varrho_0^2 - 2\varrho_0\varrho \cos(\theta - \theta_0)]} = r. \qquad (2)$$

This is the required equation of the given circle.

For, any point M which lies on the given circle satisfies the condition (1) and consequently the coordinates of M will satisfy the equation (2). But any point M which does not lie on the given circle will not satisfy the condition (1) and consequently its coordinates will not satisfy the equation (2).

The problem is therefore solved, except that we can rewrite the equation in a slightly simpler form free from radicals:

$$\varrho^2 - 2\varrho_0\varrho \cos(\theta - \theta_0) = r^2 - \varrho_0^2.$$

174. Derive the equation of the locus of points equidistant from the coordinate axes.

175. Derive the equation of the locus of points at distance a from the y-axis.

176. Derive the equation of the locus of points at distance b from the x-axis.

177. From the point $P(6, -8)$ rays are drawn in all possible directions to intersect the x-axis. Set up an equation for the locus of the mid-points of the intervals cut off on these rays between P and the x-axis.

178. From the point $C(10, -3)$ rays are drawn in all possible directions to intersect the y-axis. Set up an equation for the locus of the mid-points of the intervals cut off on these rays between C and the y-axis.

179. Derive an equation for the path of a point which moves so that it is always equidistant from the points:

1) $A(3, 2)$, $B(2, 3)$ 2) $A(5, -1)$, $B(1, -5)$,
3) $A(5, -2)$, $B(-3, -2)$ 4) $A(3, -1)$, $B(3, 5)$.

180. Set up an equation for the locus of a point which moves so that the difference of the squares of its distances from the points $A(-a, 0)$ and $B(a, 0)$ is always equal to c.

181. Derive an equation for the circle having its centre at the origin and radius r.

182. Derive an equation for the circle having its centre at $C(\alpha, \beta)$ and radius r.

183. Given the equation of a circle $x^2+y^2 = 25$, set up an equation for the locus of the mid-points of the chords of this circle whose length is equal to 8.

184. Set up an equation for the locus of a point which moves so that the sum of the squares of its distances from the points $A(-3, 0)$ and $B(3, 0)$ is equal to 50.

185. $A(a, a)$, $B(-a, a)$, $C(-a, -a)$, $D(a, -a)$ are the vertices of a square. Set up an equation for the locus of a point which moves so that the sum of the squares of its distances from the sides of this square is always equal to $6a^2$.

186. Chords of the circle given by $(x-8)^2+y^2 = 64$ are drawn through the origin in all possible directions. Set up an equation for the locus of their mid-points.

187. Derive an equation for the locus of a point which moves so that the sum of its distances from the points $F_1(-3, 0)$ $F_2(3, 0)$ is always equal to 10.

188. Derive an equation for the locus of a point which moves so that the difference of its distances from the points $F_1(-5, 0)$, $F_2(5, 0)$ is always equal to 6.

189. Derive an equation for the locus of a point which moves so that its distance from the point $F(3, 0)$ is equal to its distance from the straight line given by $x+3 = 0$.

190. Derive an equation for the locus of a point which moves so that the sum of its distances from two given points $F_1(-c, 0)$, $F_2(c, 0)$ is always equal to $2a$. This locus is called an ellipse and the points F_1, F_2 are the foci of the ellipse.

Prove that the equation can be put into the form

$$\frac{x^2}{a^2}+\frac{y^2}{b^2} = 1 \quad \text{where} \quad b^2 = a^2-c^2.$$

191. Derive an equation for the locus of a point which moves so that the difference of its distances from two given points $F_1(-c, 0)$, $F_2(c, 0)$ is always equal to $2a$. This locus is called a hyperbola and the points F_1, F_2 are the foci of the hyperbola.

Prove that the equation can be put into the form

$$\frac{x^2}{a^2}-\frac{y^2}{b^2} = 1 \quad \text{where} \quad b^2 = c^2-a^2.$$

192. Derive an equation for the locus of a point which moves so that its distance from a given point $F(a, 0)$ is always equal to its distance from the straight line given by $x = -a$. This locus is called a parabola, the point F is its focus, and the given line is the directrix.

193. Derive an equation for the locus of a point which moves so that the ratio of its distance from a given point

$F(-4, 0)$ to its distance from the straight line given by $4x+25 = 0$ is always equal to 4/5.

194. Derive an equation for the locus of a point which moves so that the ratio of its distance from a given point $F(-5, 0)$ to its distance from the straight line given by $5x+16 = 0$ is always equal to 5/4.

195. Derive an equation for the locus of a point which moves so that its shortest distances from the two circles given by

$$(x+3)^2+y^2 = 1 \quad \text{and} \quad (x-3)^2+y^2 = 81$$

are always equal.

196. Derive an equation for the locus of a point which moves so that its shortest distances from the two circles given by

$$(x+10)^2+y^2 = 289 \quad \text{and} \quad (x-10)^2+y^2 = 1$$

are always equal.

197. Derive an equation for the locus of a point which moves so that its shortest distance from the circle given by

$$(x-5)^2+y^2 = 9$$

is equal to its distance from the straight line given by $x+2 = 0$.

198. Derive an equation in polar coordinates for the straight line which is perpendicular to the polar axis and which cuts off on it an interval of measure 3.

199. Derive an equation in polar coordinates for the ray which starts from the pole and makes an angle $\pi/3$ with the polar axis.

200. Derive an equation in polar coordinates for the straight line which goes through the pole and makes an angle 45° with the polar axis.

201. Set up an equation in polar coordinates for the locus of a point which moves so that its distance from the polar axis is always equal to 5.

202. Set up an equation in polar coordinates for the circle of radius $R = 5$ which goes through the pole and has its centre on the polar axis.

203. Set up an equation in polar coordinates for the circle of radius $R = 3$ which touches the polar axis at the pole.

§ 11. Parametric Equations of a Curve

Suppose that x, y are two functions of an argument t:

$$x = \varphi(t), \quad y = \psi(t). \tag{1}$$

Regard x, y as the coordinates of a point M.

As t varies, the values of x and y will, in general, also vary, and the point M will move. The equalities (1) are called the parametric equations of the curve traced out by the point M; the argument t is called the parameter. If t can be eliminated from the equalities (1), then an equation of the form $F(x, y) = 0$ is obtained for the path of M.

204. A beam AB (Fig. 8) slides along two straight lines which are taken as the coordinate axes. The point M divides the beam into two parts $AM = a$, $BM = b$. Obtain parametric equations for the locus of the point M, taking as parameter $t = \angle OBA$. Then eliminate the parameter t to obtain an equation for the locus of M in the form $F(x, y) = 0$.

205. The locus of a point M is the ellipse given by $\frac{x^2}{a^2}+\frac{y^2}{b^2}=1$ (see q. **190**). Obtain parametric equations for the path of M, taking as parameter t the angle of inclination of the interval $\overline{OM}$ to the x-axis.

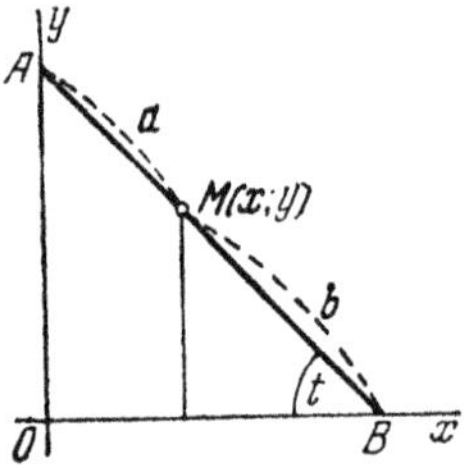

FIG. 8

206. The locus of a point M is the hyperbola given by $\frac{x^2}{a^2}-\frac{y^2}{b^2}=1$ (see q. **191**). Obtain parametric equations for the path of M, taking as parameter t the angle of inclination of the interval $\overline{OM}$ to the x-axis.

207. The locus of a point M is the parabola given by $y^2 = 4ax$ (see q. **192**). Obtain parametric equations for the path of M, taking as parameter t:

1) the y-coordinate of the point M,
2) the angle of inclination of the interval $\overline{OM}$ to the x-axis,
3) the angle of inclination of the interval $\overline{FM}$ to the x-axis, where F is the focus of the parabola.

208. The polar equations of three curves are given:

$$1)\ \varrho = 2R\cos\theta \quad 2)\ \varrho = 2R\sin\theta \quad 3)\ \varrho = 2p\,\frac{\cos\theta}{\sin^2\theta}.$$

Set up parametric equations for these curves in a rectangular Cartesian coordinate system whose origin coincides with the pole and which has the positive x-axis along the polar axis, taking the polar angle as parameter.

209. Eliminate the parameter t and obtain an equation in the form $F(x, y) = 0$ for each of the following curves given by their parametric equations:

1) $\left.\begin{array}{l} x = t^2 - 2t + 1 \\ y = t - 1 \end{array}\right\}$ 2) $\left.\begin{array}{l} x = a \cos t \\ y = a \sin t \end{array}\right\}$ 3) $\left.\begin{array}{l} x = a \sec t \\ y = b \tan t \end{array}\right\}$

4) $\left.\begin{array}{l} x = (t + 1/t)a/2 \\ y = (t - 1/t)b/2 \end{array}\right\}$ 5) $\left.\begin{array}{l} x = 2R \cos^2 t \\ y = R \sin 2t \end{array}\right\}$ 6) $\left.\begin{array}{l} x = R \sin 2t \\ y = 2R \sin^2 t \end{array}\right\}$

7) $\left.\begin{array}{l} x = 2p \cot^2 t \\ y = 2p \cot t \end{array}\right\}$

CHAPTER 3

CURVES OF THE FIRST ORDER

§ 12. General Equation of a Straight Line. Equation of a Straight Line in Terms of its Gradient. Angle between Two Straight Lines. Condition for Two Lines to be Parallel or to be Perpendicular

In Cartesian coordinates any straight line is defined by an equation of the first degree, and conversely any equation of the first degree defines a straight line.

An equation of the form

$$Ax+By+C=0 \tag{1}$$

is called the general equation of a straight line.

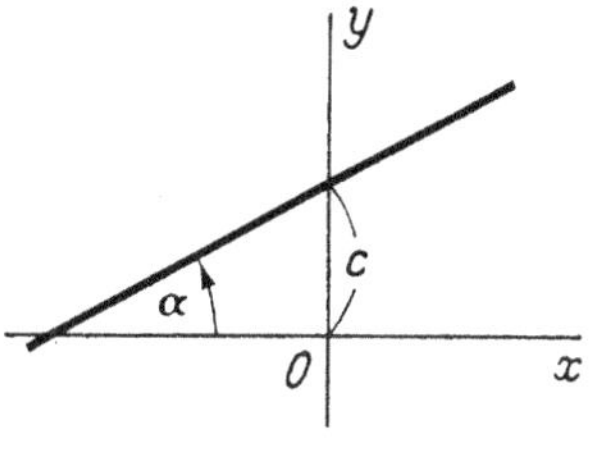

FIG. 9

The angle α, defined as indicated in Fig. 9, is called the angle of inclination of the straight line to the x-axis. The tangent of the angle of inclination of a straight line to the x-axis is called the gradient of the straight line, and it is usually denoted by the letter m:

$$m = \tan \alpha.$$

The equation $y = mx + c$ is the equation of a straight line in terms of its gradient: m is the gradient, and c is the measure of the interval cut off by the straight line on the y-axis, measured from the origin.

The gradient of a straight line given by a general equation

$$Ax + By + C = 0$$

is $m = -A/B$.

The equation of a straight line which passes through the point $M_0(x_0, y_0)$ and has gradient m is

$$y - y_0 = m(x - x_0).$$

The gradient of the straight line which passes through the points $M_1(x_1, y_1)$, $M_2(x_2, y_2)$ is given by

$$m = \frac{y_2 - y_1}{x_2 - x_1}.$$

The equation of the straight line which passes through the points $M_1(x_1, y_1)$, $M_2(x_2, y_2)$ is

$$\frac{x - x_1}{x_2 - x_1} = \frac{y - y_1}{y_2 - y_1}.$$

The angle φ between two straight lines of known gradients m_1, m_2 is given by

$$\tan \varphi = \frac{m_2 - m_1}{1 + m_1 m_2}.$$

Two lines will be parallel if they have equal gradients

$$m_1 = m_2.$$

Two lines will be perpendicular if they have gradients of opposite sign which are reciprocals in absolute value, i.e., if

$$m_1 m_2 = -1 \quad \text{or} \quad m_1 = -1/m_2.$$

210. Which of the points $M_1(3, 1)$, $M_2(2, 3)$, $M_3(6, 3)$, $M_4(-3, -3)$, $M_5(3, -1)$, $M_6(-2, 1)$ lie on the straight line given by $2x-3y-3=0$?

211. The points P_1, P_2, P_3, P_4 lie on the straight line given by $3x-2y-6=0$, and their x-coordinates are respectively 4, 0, 2, -2. Find their y-coordinates.

212. The points Q_1, Q_2, Q_3, Q_4, Q_5 lies on the straight line given by $x-3y+2=0$, and their y-coordinates are respectively 1, 0, 2, -1, 3. Find their x-coordinates.

213. Find the points of intersection with the coordinate axes of the straight line given by $2x-3y-12=0$, and draw its graph.

214. Find the point of intersection of the two straight lines given by $3x-4y-29=0$, $2x+5y+19=0$.

215. The sides AB, BC, AC of the triangle ABC are given respectively by the equations†

$$4x+3y-5=0, \quad x-3y+10=0, \quad x-2=0.$$

Find the coordinates of the vertices of the triangle.

216. Two sides of a parallelogram are given by

$$8x+3y+1=0, \quad 2x+y-1=0$$

and one of its diagonals is given by $3x+2y+3=0$. Find the coordinates of the vertices of this parallelogram.

217. The sides of a triangle lie on the straight lines given by

$$x+5y-7=0, \quad 3x-2y-4=0, \quad 7x+y+19=0.$$

Calculate the area S of this triangle.

† By the "equations of the sides" we mean, here and in the rest of the book, the equations of the straight lines on which the sides lie.

218. The area S of a triangle $= 8$ units of area. Two of its vertices are the points $A(1, -2)$ and $B(2, 3)$, and the third vertex C lies on the straight line given by $2x+y-2=0$. Find the coordinates of the vertex C.

219. The area S of a triangle $= 1{\cdot}5$ units of area. Two of its vertices are the points $A(2, -3)$ and $B(3, -2)$, and its centre of gravity lies on the straight line given by $3x-y-8=0$. Find the coordinates of its third vertex C.

220. Write down the equation, and draw the graph, of the straight line whose gradient is m and which makes an intercept c on the y-axis if

1) $m = 2/3$, $c = 3$, 2) $m = 3$, $c = 0$,
3) $m = 0$, $c = -2$, 4) $m = -3/4$, $c = 3$,
5) $m = -2$, $c = -5$, 6) $m = -1/3$, $c = 2/3$.

221. Find the gradient m and the intercept c on the y-axis for the straight line given by:

1) $5x-y+3=0$, 2) $2x+3y-6=0$, 3) $5x+3y+2=0$
4) $3x+2y=0$, 5) $y-3=0$.

222. Find the gradient m of a straight line which is (1) parallel, (2) perpendicular, to the straight line given by $5x+3y-3=0$.

223. Write down the equation of a straight line which goes through the point $M_0(2, 1)$ and is (1) parallel, (2) perpendicular, to the straight line given by $2x+3y+4=0$.

224. Given the equations of two sides of a rectangle

$$2x-3y+5=0, \quad 3x+2y-7=0$$

and one of its vertices $A(2, -3)$, find the equations of its other two sides.

225. Find the vertices of a rectangle two of whose sides are given by

$$x-2y=0, \quad x-2y+15=0$$

and one of whose diagonals is given by $7x+y-15=0$.

226. Find the projection of the point $P(-6, 4)$ on to the straight line given by $4x+5y+3=0$.

227. Find the point Q which is the mirror-image of the point $P(-5, 13)$ in the straight line given by $2x-3y-3=0$.

228. Find the equation of the straight line which is parallel to, and midway between, the straight lines given by:

1) $3x-2y-1=0, \quad 3x-2y-13=0.$
2) $5x+y+3=0, \quad 5x+y-17=0.$
3) $2x+3y-6=0, \quad 4x+6y+17=0.$
4) $5x+7y+15=0, \quad 5x+7y+3=0.$
5) $3x-15y-1=0, \quad x-5y-2=0.$

229. Calculate the gradient of the straight lines passing through the pairs of points: a) $M_1(2, -5)$, $M_2(3, 2)$, b) $P(-3, 1)$, $Q(7, 8)$, c) $A(5, -3)$, $B(-1, 6)$.

230. Set up the equations for the straight lines passing through the vertices of a triangle, $A(5, -4)$, $B(-1, 3)$, $C(-3, -2)$, which are parallel to the opposite sides.

231. $M_1(2, 1)$, $M_2(5, 3)$, $M_3(3, -4)$ are the mid-points of the sides of a triangle. Set up the equations of its sides.

232. $P(2, 3)$, $Q(-1, 0)$ are two given points. Set up the equation for the straight line which passes through the point Q and is perpendicular to the interval $\overline{PQ}$.

233. $P(2, 3)$ is the foot of the perpendicular from the origin on to a certain straight line. Set up an equation for this line.

234. Set up equations for the altitudes of a triangle whose vertices are $M_1(2, 1)$, $M_2(-1, -1)$, $M_3(3, 2)$.

235. Find the point of intersection of the altitudes of the triangle whose sides are given by $4x-y-7=0$, $x+3y-31=0$, $x+5y-7=0$.

236. $A(1, -1)$, $B(-2, 1)$, $C(3, 5)$ are the vertices of a triangle. Set up an equation for the straight line drawn through the vertex A perpendicular to the median through the vertex B.

237. $A(2, -2)$, $B(3, -5)$, $C(5, 7)$ are the vertices of a triangle. Set up an equation for the straight line drawn through the vertex C perpendicular to the internal bisector of the angle A.

238. Set up equations for the sides and medians of the triangle whose vertices are $A(3, 2)$, $B(5, -2)$, $C(1, 0)$.

239. Find the points of intersection with the coordinate axes of the straight line drawn through the points $M_1(-1, 2)$, $M_2(2, 3)$.

240. Show that the condition for the three points $M_1(x_1, y_1)$, $M_2(x_2, y_2)$, $M_3(x_3, y_3)$ to be collinear may be written in the form

$$\begin{vmatrix} x_1 & y_1 & 1 \\ x_2 & y_2 & 1 \\ x_3 & y_3 & 1 \end{vmatrix} = 0.$$

241. Show that the equation of the straight line passing through the points $M_1(x_1, y_1)$, $M_2(x_2, y_2)$ may be written in the form

$$\begin{vmatrix} x & y & 1 \\ x_1 & y_1 & 1 \\ x_2 & y_2 & 1 \end{vmatrix} = 0.$$

242. Find the point of intersection of the diagonals of the convex quadrangle whose vertices are the points $A(-3, 1)$, $B(3, 9)$, $C(7, 6)$, $D(-2, -6)$.

243. Set up equations for the sides of the parallelogram $ABCD$ which has two adjacent vertices at $A(-3, -1)$ and $B(2, 2)$ and whose diagonals intersect in the point $Q(3, 0)$.

244. Given the equations of two sides of a rectangle

$$5x+2y-7 = 0, \quad 5x+2y-36 = 0$$

and the equation of one of its diagonals $3x+7y-10 = 0$, set up equations for its remaining sides and the other diagonal.

245. Set up equations for the internal and external bisectors of the angle A of the triangle whose vertices are the points $A(1, -2)$, $B(5, 4)$, $C(-2, 0)$.

246. Set up an equation for the straight line passing through the point $P(3, 5)$ and equidistant from the points $A(-7, 3)$, $B(11, -15)$.

247. Find the projection of the point $P(-8, 12)$ on to the straight line passing through the points $A(2, -3)$, $B(-5, 1)$.

248. Find the point M_1 which is the mirror-image of the point $M_2(8, -9)$ in the line passing through the points $A(3, -4)$, $B(-1, -2)$.

249. Find a point P on the x-axis such that the sum of its distances from the points $M(1, 2)$ and $N(3, 4)$ is a minimum.

250. Find a point P on the y-axis such that the difference of its distances from the points $M(-3, 2)$ and $N(2, 5)$ is a maximum.

251. Find a point P on the straight line given by $2x-y-5=0$ such that the sum of its distances from the two points $A(-7, 1)$, $B(-5, 5)$ is a minimum.

252. Find a point P on the straight line given by $3x-y-1=0$ such that the difference of its distances from the two points $A(4, 1)$ and $B(0, 4)$ is a maximum.

253. Determine the angle φ between the two lines given by:

1) $5x-y+7=0,\quad 3x+2y=0$
2) $3x-2y+7=0,\quad 2x+3y-3=0$
3) $x-2y-4=0,\quad 2x-4y+3=0$
4) $3x+2y-1=0,\quad 5x-2y+3=0.$

254. Set up an equation for the straight line which passes through the point $M_0(2, 1)$ and makes an angle of 45° with the straight line given by $2x+3y+4=0$.

255. The point $A(-4, 5)$ is the vertex of a square whose diagonal lies on the line given by $7x-y+8=0$. Set up equations for the sides and the other diagonal of this square.

256. Set up equations for the sides of a square which has the points $A(-1, 3)$ and $C(6, 2)$ as opposite vertices.

257. The point $E(1, -1)$ is the centre of a square one of whose sides lies on the straight line given by $x-2y+12=0$. Set up equations for the straight lines on which the other sides lie.

258. A ray of light is directed from the point $M_0(-2, 3)$ to make an angle α with the x-axis and is reflected from the x-axis. It is known that $\tan\alpha = 3$. Set up equations for the straight lines on which the incident and reflected rays lie.

259. A ray of light is directed along the straight line given by $x-2y+5=0$ and is reflected in the straight line given by $3x-2y+7=0$. Set up an equation for the straight line on which the reflected ray lies.

260. Prove that the triangle formed by the straight lines given by

$$3x+4y-1=0, \quad x-7y-17=0, \quad 7x+y+31=0$$

is isosceles. (Compare the angles of the triangle.)

261. Prove that the equation for the straight line which goes through the point $M_1(x_1, y_1)$ and is parallel to the straight line given by

$$Ax+By+C=0$$

can be written as $A(x-x_1)+B(y-y_1)=0$.

262. Set up an equation for the straight line which goes through the point $M_1(2, -3)$ and is parallel to the straight line given by:

1) $3x-7y+3=0$, 2) $x+9y-11=0$,

3) $16x-24y-7=0$, 4) $2x+3=0$, 5) $3y-1=0$.

Solve the problem without calculating the gradients of the given lines.

Hint. Use the result of q. **261.**

263. Show that the condition for the two lines given by the equations

$$A_1x+B_1y+C_1=0 \quad \text{and} \quad A_2x+B_2y+C_2=0$$

to be perpendicular may be written in the form $A_1A_2+B_1B_2=0$.

264. Without calculating the gradients, determine which of the lines given by the following pairs of equations are perpendicular:

1) $3x-y+5=0, \; x+3y-1=0.$
2) $3x-4y+1=0, \; 4x-3y+7=0.$
3) $6x-15y+7=0, \; 10x+4y-3=0.$
4) $9x-12y+5=0, \; 8x+6y-13=0.$
5) $7x-2y+1=0, \; 4x+6y+17=0.$
6) $5x-7y+3=0, \; 3x+2y-5=0.$

Hint. Use the condition for perpendicularity obtained in q. **263.**

265. Show that the formula for determining the angle φ between the straight lines given by

$$A_1x+B_1y+C_1=0 \quad \text{and} \quad A_2x+B_2y+C_2=0$$

can be written in the form

$$\tan\varphi=\frac{A_1B_2-A_2B_1}{A_1A_2+B_1B_2}.$$

266. Without calculating the gradients, determine the angle φ between the lines given by the following pairs of equations:

1) $3x-y+5=0,\ 2x+y-7=0.$

2) $x\sqrt{2}-y\sqrt{3}-5=0,\ (3+\sqrt{2})x+(\sqrt{6}-\sqrt{3})y+7=0.$

3) $x\sqrt{3}+y\sqrt{2}-2=0,\ x\sqrt{6}-3y+3=0.$

Hint. Use the formula obtained in q. **265.**

267. $M_1(-10, 2)$ and $M_2(6, 4)$ are two vertices of a triangle whose altitudes intersect in the point $N(5, 2)$. Find the coordinates of its third vertex M_3.

268. Set up equations for the sides of a triangle which has $A(3, -1)$, $B(5, 7)$ as two of its vertices and the point $N(4, -1)$ as the point of intersection of its altitudes.

269. For a triangle ABC the equation of the side AB is $5x-3y+2=0$, the equation of the altitude AN is $4x-3y+1=0$ and of BN is $7x+2y-22=0$. Set up equations for the other two sides and for the third altitude.

270. Set up equations for the sides of a triangle ABC which has one vertex at $A(1, 3)$ if the equations of two its medians are

$$x-2y+1=0 \quad \text{and} \quad y-1=0.$$

271. Set up equations for the sides of a triangle which has one vertex at $B(-4, -5)$ if the equations of two of its altitudes are

$$5x+3y-4=0 \quad \text{and} \quad 3x+8y+13=0.$$

272. Set up equations for the sides of a triangle if it has one vertex at $A(4, -1)$ and the equations of the internal bisectors of two of its angles are $x-1=0$ and $x-y-1=0$.

273. Set up equations for the sides of a triangle, knowing one of its vertices $B(2, 6)$ and also the equation of a bisector of one of its angles $7x+y+5 = 0$ and the equation of its altitude which goes through the same angle $x-7y+15 = 0$.

274. Set up equations for the sides of a triangle, knowing one of its vertices $B(2, -1)$ and also the equation of an altitude $3x-4y+27 = 0$ and the equation of a bisector of an angle $x+2y-5 = 0$ which are drawn through different vertices.

275. Set up equations for the sides of a triangle, knowing one of its vertices $C(4, -1)$, and also the equation of an altitude $2x-3y+12=0$ and the equation of a median $2x+3y = 0$ which are drawn through the same vertex.

726. Set up equations for the sides of a triangle, knowing one of its vertices $B(2, -7)$, and also the equation of an altitude $3x+y+11 = 0$ and the equation of a median $x+2y+7 = 0$ which are drawn through different vertices.

277. Set up equations for the sides of a triangle, knowing one of its vertices $C(4, 3)$, and also the equation of a bisector of an angle $x+2y-5 = 0$ and the equation of a median $4x+13y-10 = 0$ which are drawn through the same vertex.

278. Set up equations for the sides of a triangle, knowing one of its vertices $A(3, -1)$, and also the equation of a bisector of an angle $x-4y+10 = 0$ and the equation of a median $6x+10y-59 = 0$ which are drawn through different vertices.

279. Set up an equation for a straight line which goes through the origin and which forms with the lines given by

$$x-y+12 = 0, \quad 2x+y+9 = 0$$

a triangle with area equal to 1·5 units of area.

280. Set up an equation for the straight line which goes through $P(3, 0)$ and which has cut off on it by the straight lines given by

$$2x-y-2=0, \quad x+y+3=0$$

an interval which is bisected by the point P.

281. Lines are drawn through the point $P(-3, -1)$ in all possible directions. Prove that the interval cut off on each of them by the lines given by $x-2y-3=0$, $x-2y+5=0$ is bisected by the point P.

282. Lines are drawn through the point $P(0, 1)$ in all possible directions. Prove that there is not one line such that the interval cut off on it by the lines given by $x-2y-3=0$, $x-2y+17=0$ is bisected by the point P.

283. Set up an equation for a straight line going through the origin, given that the length of the interval cut off on it by the lines given by $2x-y+5=0$, $2x-y+10=0$ is equal to $\sqrt{10}$.

284. Set up an equation for a straight line going through the point $C(-5, 4)$, given that the length of the interval cut off on it by the lines given by $x+2y+1=0$, $x+2y-1=0$ is equal to 5.

§ 13. Degenerate Equations of a Straight Line. Equation of a Straight Line in Terms of its Intercepts on the Axes. Compatibility of the Equations of Two and Three Lines

If in the general equation of a straight line

$$Ax+By+C=0 \tag{1}$$

either one or two of the coefficients A, B, C vanishes, then the equation is said to be degenerate. The following cases can occur:

1) $C=0$: the equation takes the form $Ax+By=0$ and defines a straight line going through the origin.

2) $B=0$ $(A \neq 0)$: the equation has the form $Ax+C=0$ and defines a straight line perpendicular to the x-axis. This equation can be written in the form $x=a$, where $a=-C/A$ is the measure of the interval cut off by the line on the x-axis, measured from the origin.

3) $B=0$, $C=0$ $(A \neq 0)$: the equation can be written in the form $x=0$ and defines the y-axis.

4) $A=0$ $(B \neq 0)$: the equation has the form $By+C=0$ and defines a straight line perpendicular to the y-axis. This equation can be written in the form $y=b$, where $b=-C/B$ is the measure of the interval cut off on the y-axis by the line, measured from the origin.

5) $A=0$, $C=0$ $(B \neq 0)$: the equation can be written in the form $y=0$ and defines the x-axis.

If none of the coefficients of equation (1) is zero, then the equation can be put into the form

$$\frac{x}{a}+\frac{y}{b}=1 \tag{2}$$

where $a=-C/A$ and $b=-C/B$ are the magnitudes of the intervals which the straight line cuts off on the coordinate axes. The equation (2) is the equation of a straight line in terms of its intercepts on the axes.

If two straight lines are given by the equations

$$A_1x+B_1y+C_1=0 \quad \text{and} \quad A_2x+B_2y+C_2=0,$$

then there are three possibilities:

a) $\frac{A_1}{A_2} \neq \frac{B_1}{B_2}$ – the straight lines have a single common point;

b) $\dfrac{A_1}{A_2} = \dfrac{B_1}{B_2} \neq \dfrac{C_1}{C_2}$ – the straight lines are parallel;

c) $\dfrac{A_1}{A_2} = \dfrac{B_1}{B_2} = \dfrac{C_1}{C_2}$ – the straight lines coincide, i.e., both equations define the same straight line.

285. For what values of a will the straight line given by

$$(a+2)x+(a^2-9)y+3a^2-8a+5=0$$

be (1) parallel to the x-axis?
(2) parallel to the y-axis?
(3) a line going through the origin?

Write down the equation of the line in each case.

286. For what values of m and n will the straight line given by

$$(m+2n-3)x+(2m-n+1)y+6m+9=0$$

be parallel to the x-axis and intercept on the y-axis an interval of measure -3? Write down the equation of this line.

287. For what values of m and n will the straight line given by

$$(2m-n+5)x+(m+3n-2)y+2m+7n+19=0$$

be parallel to the y-axis and intercept on the x-axis an interval of measure $+5$? Write down the equation of this line.

288. Prove that the lines given by the pair of equations intersect and find their point of intersection:

1) $x+5y-35=0,\quad 3x+2y-27=0$
2) $14x-9y-24=0\quad 7x-2y-17=0$
3) $12x+15y-8=0,\quad 16x+9y-7=0$
4) $8x-33y-19=0,\quad 12x+55y-19=0$
5) $3x+5=0,\quad y-2=0.$

289. Prove that the lines given by the pair of equations are parallel:

1) $3x+5y-4=0, \quad 6x+10y+7=0$
2) $2x-4y+3=0, \quad x-2y=0$
3) $2x-1=0, \quad x+3=0$
4) $y+3=0, \quad 5y-7=0.$

290. Prove that the lines given by the pair of equations coincide:

1) $3x+5y-4=0, \quad 6x+10y-8=0$
2) $x-y\sqrt{2}=0, \quad x\sqrt{2}-2y=0$
3) $x\sqrt{3}-1=0, \quad 3x-\sqrt{3}=0.$

291. For what values of a and b will the straight lines given by

$$ax-2y-1=0, \quad 6x-4y-b=0$$

1) have a common point 2) be parallel 3) coincide?

292. For what values of m and n will the straight lines given by

$$mx+8y+n=0, \quad 2x+my-1=0$$

1) be parallel 2) coincide 3) be perpendicular?

293. For what value of m will the straight lines given by

$$(m-1)x+my-5=0, \quad mx+(2m-1)y+7=0$$

intersect on the x-axis?

294. For what value of m will the straight lines given by

$$mx+(2m+3)y+m+6=0, \quad (2m+1)x+(m-1)y+m-2=0$$

intersect on the y-axis?

295. Do the straight lines given by the three equations intersect in the same point?

1) $2x+3y-1=0,\quad 4x-5y+5=0,\quad 3x-y+2=0$
2) $3x-y+3=0,\quad 5x+3y-7=0,\quad x-2y-4=0$
3) $2x-y+1=0,\quad x+2y-17=0,\quad x+2y-3=0.$

296. Prove that if the three straight lines given by

$$A_1x+B_1y+C_1=0,\quad A_2x+B_2y+C_2=0,\quad A_3x+B_3y+C_3=0$$

intersect in the same point, then

$$\begin{vmatrix} A_1 & B_1 & C_1 \\ A_2 & B_2 & C_2 \\ A_3 & B_3 & C_3 \end{vmatrix}=0.$$

297. Prove that if

$$\begin{vmatrix} A_1 & B_1 & C_1 \\ A_2 & B_2 & C_2 \\ A_3 & B_3 & C_3 \end{vmatrix}=0,$$

then the three straight lines given by

$$A_1x+B_1y+C_1=0,\quad A_2x+B_2y+C_2=0,$$
$$A_3x+B_3y+C_3=0$$

either intersect in the same point or are parallel or coincide.

298. For what value of a will the three straight lines given by

$$2x-y+3=0,\quad x+y+3=0,\quad ax+y-13=0$$

intersect in the same point?

299. Write the equation in terms of the intercepts on the axes and draw the graph for each of the straight lines given by

1) $2x+3y-6=0$, 2) $4x-3y+24=0$,
3) $2x+3y-9=0$, 4) $3x-5y-2=0$,
5) $5x+2y-1-0.$

300. Calculate the area of the triangle formed by the straight line given by $3x-4y-12=0$ with the coordinate axes.

301. Set up an equation for the straight line going through the point $M_1(3, -7)$ and cutting off on the axes non-zero intercepts of equal measure (measured from the origin).

302. Set up an equation for a straight line going through the point $P(2, 3)$ and cutting off on the axes intercepts of equal length.

303. Set up an equation for a straight line going through the point $C(1, 1)$ and forming with the axes a triangle with an area of 2 units of area.

304. Set up an equation for a straight line which goes through the point $B(5, -5)$ and forms with the axes a triangle with an area of 50 units of area.

305. Set up an equation for a straight line which goes through the point $P(8, 6)$ and forms with the axes a triangle with an area of 12 units of area.

306. Set up an equation for a straight line which goes throughthe point $P(12, 6)$ and forms with the axes a triangle with an area of 150 units of area.

307. A straight line is drawn through the point $M(4, 3)$ forming with the axes a triangle of area 3 units of area. Find the points of intersection with the coordinate axes.

308. Through the point $M_1(x_1, y_1)$, where $x_1y_1>0$, a straight line given by $x/a+y/b=1$ is drawn to form with the axes a triangle of area S. What relation must subsist between the quantities x_1, y_1, and S in order that the intervals a and b shall have the same sign?

§ 14. The Normal Equation of a Straight Line. How to Calculate the Distance of a Point from a Straight Line

A straight line is given in the plane xOy. The straight line drawn from the origin perpendicular to the given straight line is called its normal. Let P be the point of intersection of the normal and given straight line, and regard the direction from O to P as positive.

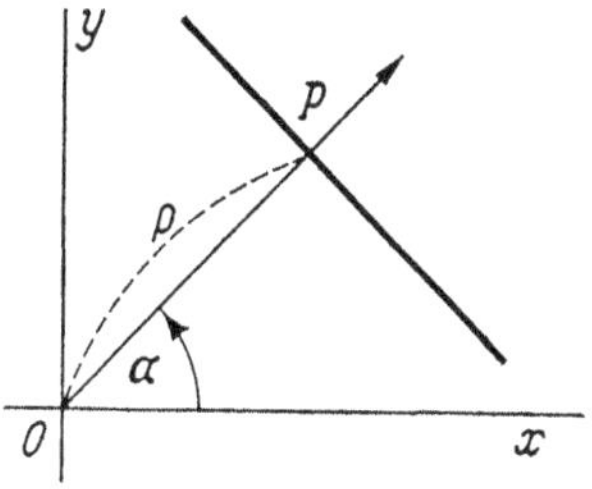

FIG. 10

The equation of the given line can be written in the form

$$x \cos \alpha + y \sin \alpha - p = 0,$$

where α is the polar angle of the normal, and p is the length of the interval $\overline{OP}$.

Let d be the distance of an arbitrary point M_0 from the given straight line. Then we call the displacement of M_0 from the given straight line:

$+d$ if M_0 and the origin lie on different sides of the given line,
$-d$ if M_0 and the origin lie on the same side of the given line, and
0 if M_0 lies on the given line.

The displacement δ of the point M_0 from the straight line given by $x \cos \alpha + y \sin \alpha - p = 0$ can be found by substituting the coordinates of M_0 for the variables in the left-hand member of the normal equation, i.e.,

$$\delta = x_0 \cos \alpha + y_0 \sin \alpha - p.$$

The distance d of the point M_0 from the given line is found by calculating δ and taking its absolute value:

$$d = |\delta|.$$

The general equation of a straight line $Ax + By + C = 0$ can be put into normal form by multiplying each term by the normalizing factor μ, defined by

$$\mu = \pm \frac{1}{\sqrt{(A^2 + B^2)}},$$

where the sign of μ is to be taken as opposite to the sign of the constant term C in the equation that is being normalized.

309. Which of the following equations of straight lines are in normal form?

1) $3x/5 - 4y/5 - 3 = 0$, 2) $2x/5 - 3y/5 - 1 = 0$
3) $5x/13 - 12y/13 + 2 = 0$ 4) $-5x/13 + 12y/13 - 2 = 0$
5) $-x + 2 = 0$ 6) $x - 2 = 0$
7) $y + 2 = 0$ 8) $-y - 2 = 0$.

310. Put the following general equations of straight lines into normal form:

1) $4x - 3y - 10 = 0$ 2) $4x/5 - 3y/5 + 10 = 0$
3) $12x - 5y + 13 = 0$ 4) $x + 2 = 0$
5) $2x - y - \sqrt{5} = 0$.

311. Find the polar angle of the normal α and the interval p for the straight lines given by the following equations. Having found α, p, draw the graph of each line (in the last two examples, take $\beta = 30°$ and $q = 2$).

1) $x-2=0$ 2) $x+2=0$ 3) $y-3=0$ 4) $y+3=0$

5) $x-\sqrt{3}+y-6=0$ 6) $x-y+2=0$ 7) $x+y\sqrt{3}+2=0$

8) $x \cos\beta - y \sin\beta - q = 0$

where $q > 0$ and β is an acute angle

9) $x \cos\beta + y \sin\beta + q = 0$

where $q > 0$ and β is an acute angle.

312. Calculate the measure of the displacement δ and the distance d of the point from the the straight line in the following cases:

1) $A(2, -1)$, $4x+3y+10=0$,

2) $B(0, -3)$, $5x-12y-23=0$,

3) $P(-2, 3)$, $3x-4y-2=0$,

4) $Q(1, -2)$, $x-2y-5=0$.

313. Do the point $M(1, --3)$ and the origin lie on the same side or different sides of the straight line defined by each of the following equations?

1) $2x-y+5=0$ 2) $x-3y-5=0$ 3) $3x+2y-1=0$

4) $x-3y+2=0$ 5) $10x+24y+15=0$.

314. The point $A(2, -5)$ is the vertex of a square one of whose sides lies on the straight line given by $x-2y-7=0$. Calculate the area of the square.

315. Calculate the area of the rectangle which has the point $A(-2, 1)$ as one vertex and two of whose sides lie on the straight lines given by

$$3x-2y-5=0 \quad \text{and} \quad 2x+3y+7=0.$$

316. Prove that the straight line given by $2x+y+3=0$ intersects the interval bounded by the points $A(-5, 1)$ and $B(3, 7)$.

317. Prove that the straight line given by $2x-3y+6=0$ does not intersect the interval bounded by the points $M_1(-2, -3)$, $M_2(1, -2)$.

318. Consecutive vertices of a quadrangle are the points $A(-3, 5)$, $B(-1, -4)$, $C(7, -1)$, $D(2, 9)$. Without plotting the points, determine whether the quadrangle is convex.

319. $A(-1, 6)$, $B(1, -3)$, $C(4, 10)$, $D(9, 0)$ are consecutive vertices of a quadrangle. Without plotting the points, determine whether the quadrangle is convex.

320. $A(-10, -13)$, $B(-2, 3)$, $C(2, 1)$ are the vertices of a triangle. Calculate the length of the perpendicular drawn from B on to the median through C.

321. The sides AB, BC, CA of a triangle ABC lie respectively on the straight lines given by

$$x+21y-22=0, \quad 5x-12y+7=0, \quad 4x-33y+146=0.$$

Calculate the distance of the centre of gravity of this triangle from the side BC.

322. Calculate the distance d between the parallel straight lines given by:

1) $3x-4y-10=0, \quad 6x-8y+5=0$
2) $5x-12y+26=0, \quad 5x-12y-13=0$
3) $4x-3y+15=0, \quad 8x-6y+25=0$
4) $24x-10y+39=0, \quad 12x-5y-26=0.$

323. Calculate the area of a square two of whose sides lie on the straight lines given by

$$5x-12y-65=0, \quad 5x-12y+26=0.$$

324. Show that the straight line given by $5x-2y-1=0$ is parallel to the straight lines given by $5x-2y+7=0$, $5x-2y-9=0$ and lies midway between them.

325. Three parallel straight lines are given by

$$10x+15y-3=0, \quad 2x+3y+5=0, \quad 2x+3y-9=0.$$

Show that the first lies between the other two and calculate the ratio in which it divides the distance between them.

326. Prove that two straight lines can be drawn through the point $P(2, 7)$ so that their distances from the point $Q(1, 2)$ are each equal to 5. Set up equations for these lines.

327. Prove that two straight lines can be drawn through the point $P(2, 5)$ so that their distances from the point $Q(5, 1)$ are each equal to 3. Set up equations for these lines.

328. Prove that only one straight line can be drawn through the point $C(7, -2)$ so that its distance from the point $A(4, -6)$ is equal to 5. Set up its equation.

329. Prove that it is impossible to draw a line through the point $B(4, -5)$ so that its distance from the point $C(-2, 3)$ is equal to 12.

330. Derive an equation for the locus of a point which moves so that its displacement from the straight line given by $8x-15y-25=0$ is equal to -2.

331. Set up equations for the straight lines parallelto, and at a distance $d = 3$ from, the straight line given by $3x-4y-10 = 0$.

332. Set up equations for the sides of a square which has the points $A(2, 0)$, $B(-1, 4)$ as two adjacent vertices.

333. The point $A(5, -1)$ is a vertex of a square one of whose sides lies on the straight line given by $4x-3y-7 = 0$. Set up equations for the straight lines on which its other sides lie.

334. $4x-3y+3 = 0$, $4x-3y-17 = 0$ are equations for two sides of a square and one of its vertices is the point $A(2, -3)$. Set up equations for its other two sides.

335. $5x+12y-10 = 0$, $5x+12y+29 = 0$ are equations for two sides of a square. Set up equations for its other two sides, given that the point $M_1(-3, 5)$ lies on one of the sides of this square.

336. Find the coordinates of a point M whose displacements from the straight lines given by $5x-12y-13 = 0$ and $3x-4y-19 = 0$ are respectively -3 and -5.

337. Set up an equation for a straight line which goes through the point $P(-2, 3)$ and is equidistant from the points $A(5, -1)$ and $B(3, 7)$.

338. Set up an equation for the locus of a point which moves so that it is equidistant from the parallel straight lines given by:

1) $3x-y+7 = 0, \quad 3x-y-3 = 0$
2) $x-2y+3 = 0, \quad x-2y+7 = 0$
3) $5x-2y-6 = 0, \quad 10x-4y+3 = 0.$

339. Set up equations for the bisectors of the angles formed by the two intersecting straight lines given by:

1) $x-3y+5=0, \quad 3x-y-2=0$
2) $x-2y-3=0, \quad 2x+4y+7=0$
3) $3x+4y-1=0, \quad 5x+12y-2=0.$

340. Set up equations for the straight lines which go through the point $P(2, -1)$ and form an isosceles triangle with the straight lines given by

$$2x-y+5=0, \quad 3x+6y-1=0.$$

341. Determine whether the point $M(1, -2)$ and the origin lie in the same angle, in adjacent angles, or in vertically opposite angles, formed by the straight lines given by:

1) $2x-y-5=0, \quad 3x+y+10=0$
2) $4x+3y-10=0, \quad 12x-5y-5=0$
3) $x-2y-1=0, \quad 3x-y-2=0.$

342. Determine whether the points $M(2, 3)$ and $N(5, -1)$ lie in the same angle, in adjacent angles, or in vertically opposite angles, formed by the straight lines given by:

1) $x-3y-5=0, \quad 2x+9y-2=0$
2) $2x+7y-5=0, \quad x+3y+7=0$
3) $12x+y-1=0, \quad 13x+2y-5=0.$

343. Does the origin lie inside or outside the triangle formed by the straight lines given by

$$7x-5y-11=0, \quad 8x+3y+31=0, \quad x+8y-19=0?$$

344. Does the point $M(-3, 2)$ lie inside or outside the triangle formed by the straight lines given by

$$x+y-4=0, \quad 3x-7y+8=0, \quad 4x-y-31=0?$$

345. Is it the acute angle or the obtuse angle formed by the straight lines given by $3x-2y+5=0$, $2x+y-3=0$ which contains the origin?

346. Is it the acute angle or the obtuse angle formed by the straight lines given by $3x-5y-4=0$, $x+2y+3=0$ which contains the point $M(2, -5)$?

347. Two straight lines are given by $3x-y-4=0$, $2x+6y+3=0$. Set up an equation for that bisector of the angle between them which passes through the origin.

348. Two straight lines are given by $x-7y+5=0$, $5x+5y-3=0$. Set up an equation for the bisector of that angle between them which is adjacent to the one containing the origin.

349. Two straight lines are given by $x+2y-11=0$, $3x-6y-5=0$. Set up an equation for the bisector of that angle between them in which the point $M(1, -3)$ lies.

350. Two straight lines are given by $2x-3y-5=0$, $6x-4y+7=0$. Set up an equation for the bisector of that angle between them which is adjacent to the one containing the point $C(2, -1)$.

351. Set up an equation for the bisector of the acute angle between the straight lines given by $3x+4y-5=0$, $5x-12y+3=0$.

352. Set up an equation for the bisector of the obtuse angle between the straight lines given by $x-3y+5=0$, $3x-y+15=0$.

§ 15. Equation of a Pencil of Straight Lines

The aggregate of straight lines which pass through a certain point S is called a pencil of straight lines with centre S.

If $A_1x+B_1y+C_1=0$ and $A_2x+B_2y+C_2=0$ are the equations of two straight lines intersecting in a point S, then the equation

$$\alpha(A_1x+B_1y+C_1)+\beta(A_2x+B_2y+C_2)=0, \tag{1}$$

where α, β may be any two numbers, not simultaneously zero, defines a straight line which also passes through the point S.

Moreover, it is always possible to choose the numbers α, β so that the equation (1) fits any previously defined straight line passing through the point S, i.e., any straight line of the pencil with centre S. The equation (1) is therefore called the equation of the pencil (with centre S).

If $\alpha \neq 0$, then on dividing both members of equation (1) by α and putting $\beta/\alpha = \lambda$, we get

$$A_1x+B_1y+C_1+\lambda(A_2x+B_2y+C_2)=0. \tag{2}$$

An equation of the form (2) can define, for varying λ, any line of the pencil with centre S except one, namely that corresponding to $\alpha=0$, i.e., except the straight line given by $A_2x+B_2y+C_2=0$.

353. Find the centre of the pencil of lines given by

$$\alpha(2x+3y-1)+\beta(x-2y-4)=0.$$

354. Find an equation for the straight line belonging to the pencil given by $\alpha(x+2y-5)+\beta(3x-2y+1)=0$ which

1) goes through the point $A(3, -1)$,
2) goes through the origin,
3) is parallel to the x-axis,
4) is parallel to the y-axis,
5) is parallel to the straight line given by $4x+3y-5=0$,
6) is perpendicular to the straight line given by $2x+3y+7=0$.

355. Set up an equation for the straight line which passes through the point of intersection of the straight lines given by

$$3x-2y+5=0, \quad 4x+3y-1=0$$

and which cuts off on the y-axis an intercept $c=-3$. Solve the problem without finding the point of intersection of the given lines.

356. Set up an equation for the straight line which passes through the point of intersection of the straight lines given by

$$2x+y-2=0, \quad x-5y-23=0$$

and which bisects the interval bounded by the points $M_1(5, -6)$, $M_2(-1, -4)$. Solve the problem without finding the point of intersection of the given straight lines.

357. A pencil of straight lines is given by

$$\alpha(3x-4y-3)+\beta(2x+3y-1)=0.$$

Write down an equation for the straight line of this pencil which goes through the centre of gravity of a uniform triangular lamina whose vertices are the points $A(-1, 2)$, $B(4, -4)$, $C(6, -1)$.

358. A pencil of straight lines is given by

$$\alpha(3x-2y-1)+\beta(4x-5y+8)=0.$$

Find an equation for the straight line of this pencil which bisects the interval cut off on the straight line given by $x+2y+$ $+4=0$ by the two straight lines given by $2x+3y+5=0$ and $x+7y-1=0$.

359. Without finding the vertices of the triangle formed by the straight lines given by

$$x+2y-1=0, \quad 5x+4y-17=0, \quad x-4y+11=0,$$

set up equations for the altitudes of this triangle.

360. Set up an equation for a straight line which passes through the point of intersection of the straight lines given by

$$2x+7y-8=0, \quad 3x+2y+5=0$$

and which makes an angle of 45° with the straight line given by $2x+3y-7=0$. Solve the problem without finding the point of intersection of the two given lines.

361. In a triangle ABC the equations of the altitudes AN, BN are respectively $x+5y-3=0$, $x+y-1=0$, and that of the side AB is $x+3y-1=0$. Without finding the vertices of the triangle or the point of intersection of the altitudes, set up equations for the other two sides and the third altitude of the triangle.

362. Set up equations for the sides of the triangle ABC, knowing one of its vertices $A(2, -1)$ and also the equation of an altitude

$$7x-10y+1=0$$

and the equation of a bisector of an angle $3x-2y+5=0$, both of these going through the same vertex. Solve the problem without calculating the coordinates of the vertices B and C.

363. A pencil of straight lines is given by

$$\alpha(2x+y+8)+\beta(x+y+3)=0.$$

Find an equation for a straight line of this pencil on which the straight lines given by $x-y-5=0$, $x-y-2=0$ cut off an interval equal to $\sqrt{5}$.

364. Show that the straight line given by $x+3y+13=0$ is a member of the pencil of straight lines given by

$$\alpha(3x+y-1)+\beta(2x-y\quad 9)=0.$$

365. Show that the straight line given by $7x+2y-15 = 0$ is not a member of the pencil of straight lines given by

$$\alpha(5x+3y+6)+\beta(3x-4y-37) = 0.$$

366. For what value of C will the straight line given by $4x-3y+C=0$ be a member of the pencil of straight lines given by

$$\alpha(3x+2y-9)+\beta(2x+5y+5) = 0?$$

367. For what values of a will the straight line given by $ax+5y+9=0$ not be a member of the pencil of straight lines given by

$$\alpha(5x+3y-7)+\beta(3x+10y+4) = 0?$$

368. The centre of the pencil of straight lines given by

$$\alpha(2x-3y+20)+\beta(3x+5y-27) = 0$$

is a vertex of a square, one of whose diagonals lies on the straight line given by $x+7y-16=0$. Set up equations for the sides and the other diagonal of this square.

369. A pencil of straight lines is given by

$$\alpha(2x+5y+4)+\beta(3x-2y+25) = 0.$$

Find the equation for the straight line of this pencil which cuts off on the coordinate axes intercepts of equal measure (non-zero and measured from the origin).

370. A pencil of straight lines is given by

$$\alpha(2x+y+1)+\beta(x-3y-10) = 0.$$

Find equations for the straight lines of this pencil which cut off on the coordinate axes intercepts of equal length (measured from the origin).

371. A pencil of straight lines is given by

$$\alpha(21x+8y-18)+\beta(11x+3y+12) = 0.$$

Find equations for the straight lines of this pencil which form with the coordinate axes triangles with area equal to 9 units of area.

372. A pencil of straight lines is given by

$$\alpha(2x+y+4)+\beta(x-2y-3) = 0.$$

Show that there is only one straight line of this pencil whose distance from the point $P(2, -3)$ is equal to $\sqrt{10}$. Write down the equation of this straight line.

373. Show that there is no straight line belonging to the pencil of lines given by $\alpha(2x-y-6)+\beta(x-y-4) = 0$ whose distance from the point $P(3, -1)$ is equal to 3.

374. Set up an equation for a straight line which passes through the point of intersection of the straight lines given by

$$3x+y-5 = 0, \quad x-2y+10 = 0$$

and which is at distance $d = 5$ from the point $C(-1, -2)$. Solve the problem without finding the point of intersection of the two given straight lines.

375. A pencil of straight lines is given by

$$\alpha(5x+2y+4)+\beta(x+9y-25) = 0.$$

Write down equations for the straight lines of this pencil which form isosceles triangles with the straight lines given by $2x-3y+5 = 0$ and $12x+8y-7 = 0$.

376. Set up an equation for a straight line which passes through the point of intersection of the straight lines given by

$$11x+3y-7 = 0, \quad 12x+y-19 = 0$$

and which is equidistant from the points $A(3, -2)$ and $B(-1, 6)$. Solve the problem without finding the point of intersection of the two given straight lines.

377.
$$\alpha_1(5x+3y-2)+\beta_1(3x-y-4)=0$$
$$\alpha_2(x-y+1)\quad+\beta_2(2x-y-2)=0$$

are the equations of two pencils of lines. Without finding the centres of these pencils, set up an equation for the straight line which belongs to both pencils.

378. The sides AB, BC, CD, DA of a quadrilateral $ABCD$ lie on the straight lines given respectively by

$$5x+y+13=0,\quad 2x-7y-17=0,\quad 3x+2y-13=0,$$
$$3x-4y+17=0.$$

Without finding the vertices of this quadrilateral, set up equations for its diagonals.

379. The centre of the pencil of straight lines given by

$$\alpha(2x+3y+5)+\beta(3x-y+2)=0$$

is one vertex of a triangle two of whose altitudes are given by

$$x-4y+1=0,\quad 2x+y+1=0.$$

Set up equations for the sides of this triangle.

§ 16. The Polar Equation of a Straight Line

A straight line drawn through the pole perpendicular to a given straight line is called the normal to the given straight line. Let P be the point of intersection of the normal with the given straight line, and regard the direction from the pole O to the point P as the positive direction of the normal. The polar angle of the normal is then the angle through which the polar axis must be rotated to make it coincide with the directed interval $\overline{OP}$ (Fig. 11).

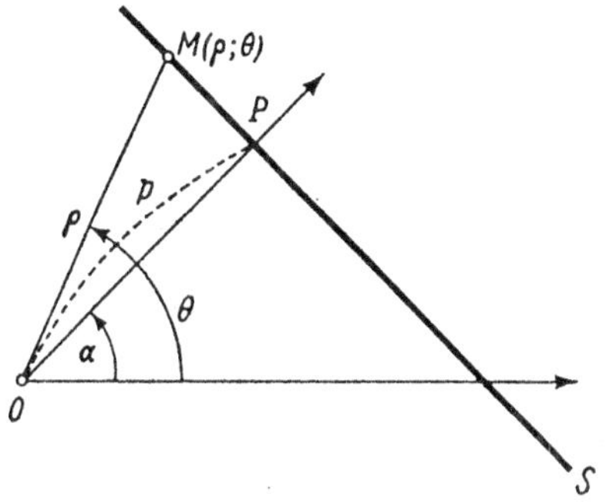

FIG. 11

380. Derive the polar equation of a straight line, knowing its distance p from the pole and the polar angle α of its normal.

Solution. 1st method.

Let M with polar coordinates ϱ, θ be any point on the given straight line s, and let P be its point of intersection with the normal. From the right-angled triangle OPM we have

$$\varrho = \frac{p}{\cos(\theta-\alpha)}. \tag{1}$$

(1) is an equation in the two variables ϱ, θ which is satisfied by the coordinates of any point M which lies on the straight line s and which is not satisfied by the coordinates of any point that does not lie on s. Hence (1) is the required equation of the given straight line s and the problem is solved.

2nd method.

The equation of the straight line s in a rectangular Cartesian coordinate system having its origin at the pole O and its x-axis along the polar axis is

$$x\cos\alpha+y\sin\alpha-p=0. \tag{2}$$

The formulae for transformation from Cartesian coordinates to polars are

$$x=\varrho\cos\theta,\quad y=\varrho\sin\theta. \tag{3}$$

Substituting for x and y in (2), we get

$$\varrho\,(\cos\theta\cos\alpha+\sin\theta\sin\alpha)=p$$

or

$$\varrho=\frac{p}{\cos(\theta-\alpha)}.$$

381. Derive the polar equation of a straight line, given

1) the angle of inclination β of the straight line to the polar axis and the length p of the perpendicular from the pole on to the given line. Write down the equation for this line when $\beta=\pi/6$, $p=3$.

2) the interval a (measured from the pole) which the straight line cuts off on the polar axis and the polar angle α of the normal to the line. Write down the equation for this line when $a=2$, $\alpha=-2\pi/3$.

3) the angle of inclination β of the straight line to the polar axis and the interval a (measured from the pole) which the line cuts off on the polar axis. Write down the equation for this line when $\beta=\pi/6$, $a=6$.

382. Derive the polar equation of a straight line which passes through the point $M_1(\varrho_1, \theta_1)$ and makes an angle β with the polar axis.

383. Derive the polar equation of a straight line which passes through the point $M_1(\varrho_1, \theta_1)$ if the polar angle of its normal is α.

384. Set up the polar equation of a straight line which passes through the points $M_1(\varrho_1, \theta_1)$ and $M_2(\varrho_2, \theta_2)$.

CHAPTER 4

GEOMETRICAL PROPERTIES OF CURVES OF THE SECOND ORDER

§ 17. The Circle

The equation $$(x-x_0)^2+(y-y_0)^2 = R^2 \tag{1}$$

defines a circle of radius R with centre $C(x_0, y_0)$.

If the centre of the circle is at the origin, i.e., if $x_0 = 0$, $y_0 = 0$, then equation (1) takes the form

$$x^2+y^2 = R^2. \tag{2}$$

385. Set up equations for the following circles:

1) centre at origin and radius $R = 3$
2) centre at $C(2, -3)$ and radius $R = 7$
3) centre at $C(6, -8)$ and circle goes through the origin
4) centre at $C(-1, 2)$ and circle goes through the point $A(2, 6)$
5) the point $A(3, 2)$, $B(-1, 6)$ are the ends of a diameter of the circle
6) centre at origin and the line given by $3x-4y+20 = 0$ is a tangent to the circle
7) centre at $C(1, -1)$ and the line given by $5x-12y+9 = 0$ touches the circle
8) the circle goes through the points $A(3, 1)$, $B(-1, 3)$ and its centre lies on the straight line given by $3x-y-2 = 0$

9) the circle goes through the three points $A(1,1)$, $B(1, -1)$, $C(2, 0)$

10) the circle goes through the three points $M_1(-1, 5)$, $M_2(-2, -2)$, $M_3(5, 5)$.

386. Set up an equation for a circle whose centre is at $C(3, -1)$ and which has a chord of length 6 lying on the straight line given by $2x-5y+18=0$.

387. Write down equations for the circles of radius $R=\sqrt{5}$ which touch the straight line given by $x-2y-1=0$ at the point $M_1(3, 1)$.

388. Set up an equation for the circle which touches the two parallel straight lines given by $2x+y-5=0$, $2x+y+15=0$, its point of contact with one of them being the point $A(2, 1)$.

389. Set up equations for the circles which go through the point $A(1, 0)$ and touch the parallel straight lines given by $2x+y+2=0$, $2x+y-18=0$.

390. Set up an equation for the circle whose centre lies on the straight line given by $2x+y=0$ and which touches the two straight lines given by $4x-3y+10=0$, $4x-3y-30=0$.

391. Set up equations for the circles which touch the two intersecting straight lines given by $7x-y-5=0$, $x+y+13=0$, its point of contact with one of them being the point $M_1(1, 2)$.

392. Set up equations for the circles which go through the origin and touch the two intersecting straight lines given by

$$x+2y-9=0, \quad 2x-y+2=0.$$

393. Set up equations for the circles whose centres lie on the straight line given by $4x-5y-3=0$ and which touch the straight lines given by $2x-3y-10=0$, $3x-2y+5=0$.

394. Write down equations for the circles which go through the point $A(-1, 5)$ and touch the two intersecting straight lines given by

$$3x+4y-35=0, \quad 4x+3y+14=0.$$

395. Write down equations for the circles which touch the three straight lines given by $4x-3y-10=0$, $3x-4y-5=0$, $3x-4y-15=0$.

396. Write down equations for the circles which touch the three straight lines given by $3x+4y-35=0$, $3x-4y-35=0$, $x-1=0$.

397. Which of the following equations define a circle? Give the centre C and radius R of each circle.

1) $(x-5)^2+(y+2)^2=25$ 2) $(x+2)^2+y^2=64$
3) $(x-5)^2+(y+2)^2=0$ 4) $x^2+(y-5)^2=5$
5) $x^2+y^2-2x+4y-20=0$ 6) $x^2+y^2-2x+4y+14=0$
7) $x^2+y^2+4x-2y+5=0$ 8) $x^2+y^2+x=0$
9) $x^2+y^2+6x-4y+14=0$ 10) $x^2+y^2+y=0$

398. What sort of curves do the following equations define? Draw their graphs.

1) $y=+\sqrt{(9-x^2)}$ 6) $y=15-\sqrt{(64-x^2)}$
2) $y=-\sqrt{(25-x^2)}$ 7) $x=-2-\sqrt{(9-y^2)}$
3) $x=-\sqrt{(4-y^2)}$ 8) $x=-2+\sqrt{(9-y^2)}$
4) $x=+\sqrt{(16-y^2)}$ 9) $y=-3-\sqrt{(21-4x-x^2)}$
5) $y=15+\sqrt{(64-x^2)}$ 10) $x=-5+\sqrt{(40-6y-y^2)}$

399. Is the point $A(1, -2)$ inside, outside, or on the circumference of, the circles given by the following equations?

1) $x^2+y^2=1$ 2) $x^2+y^2=5$ 3) $x^2+y^2=9$
4) $x^2+y^2-8x-4y-5=0$ 5) $x^2+y^2-10x+8y=0$.

400. Find an equation for the line of centres of the two circles given by:

1) $(x-3)^2+y^2=9$, $(x+2)^2+(y-1)^2=1$
2) $(x+2)^2+(y-1)^2=16$, $(x+2)^2+(y+5)^2=25$
3) $x^2+y^2-4x+6y=0$, $x^2+y^2-6x=0$
4) $x^2+y^2-x+2y=0$, $x^2+y^2+5x+2y-1=0$.

401. Set up an equation for the diameter of the circle given by $x^2+y^2+4x-6y-17=0$ which is perpendicular to the straight line given by $5x+2y-13=0$.

402. Calculate the shortest distance from the given point to the given circle in the following cases:

a) $A(6, -8)$, $x^2+y^2=9$
b) $B(3, 9)$, $x^2+y^2-26x+30y+313=0$
c) $C(-7, 2)$, $x^2+y^2-10x-14y-151=0$

403. Find the points of intersection of the straight line and circle given by $7x-y+12=0$ and $(x-2)^2+(y-1)^2=25$.

404. How are the straight line and circle given by the following equations situated? Does the straight line intersect, touch, or pass outside, the circle?

1) $y=2x-3$, $x^2+y^2-3x+2y-3=0$
2) $y=x/2-1/2$, $x^2+y^2-8x+2y+12=0$
3) $y=x+10$, $x^2+y^2-1=0$.

405. For what values of the gradient m will the straight line given by $y = mx$ (1) intersect the circle given by $x^2+y^2-10x=0$, (2) touch this circle, (3) pass outside it?

406. Derive the condition that the straight line given by $y = mx+c$ shall touch the circle given by $x^2+y^2 = R^2$.

407. Set up an equation for the diameter of the circle given by $(x-2)^2+(y+1)^2 = 16$ if the diameter goes through the mid-point of the chord of the circle which lies on the straight line given by $x-2y-3 = 0$.

408. The point $M(8{\cdot}5, 3{\cdot}5)$ is the mid-point of a chord of the circle given by $(x-3)^2+(y-7)^2 = 169$. Set up an equation for this chord.

409. The point $A(1, 2)$ is the mid-point of a chord of the circle given by $(x-2)^2+(y-4)^2 = 10$. What is the length of the chord?

410. A pencil of straight lines is given by

$$\alpha(x-8y+30)+\beta(x+5y-22) = 0.$$

Find equations for the straight lines of this pencil on which the circle given by $x^2+y^2-2x+2y-14 = 0$ intercepts chords of length $2\sqrt{3}$.

411. Two circles which intersect in the points $M_1(x_1, y_1)$, $M_2(x_2, y_2)$ are given by $(x-m_1)^2+(y-n_1)^2 = R_1^2$, $(x-m_2)^2+(y-n_2)^2 = R_2^2$. Show that any circle passing through the two points M_1, M_2, and also the straight line M_1M_2 itself, can be defined, with a suitable choice of the numbers α, β, by an equation of the form

$$\alpha[(x-m_1)^2+(y-n_1)^2-R_1^2]+\beta[(x-m_2)^2+(y-n_2)^2-R_2^2] = 0.$$

412. Set up an equation for the circle which goes through the point $A(1, -1)$ and the points of intersection of the two circles given by

$$x^2+y^2+2x-2y-23=0, \quad x^2+y^2-6x+12y-35=0.$$

413. Set up an equation for the circle which goes through the origin and the points of intersection of the two circles given by

$$(x+3)^2+(y+1)^2=25, \quad (x-2)^2+(y+4)^2=9.$$

414. Set up an equation for the straight line passing through the points of intersection of the circles given by

$$x^2+y^2+3x-y=0, \quad 3x^2+3y^2+2x+y=0.$$

415. Calculate the distance from the centre of the circle given by $x^2+y^2=2x$ to the straight line through the points of intersection of the circles given by

$$x^2+y^2+5x-8y+1=0, \quad x^2+y^2-3x+7y-25=0.$$

416. Find the length of the common chord of the two circles given by

$$x^2+y^2-10x-10y=0, \quad x^2+y^2+6x+2y-40=0.$$

417. Set up an equation for the circle whose centre lies on the straight line given by $x+y=0$ and which passes through the points of intersection of the two circles given by

$$(x-1)^2+(y+5)^2=50, \quad (x+1)^2+(y+1)^2=10.$$

418. Set up an equation for the tangent at the point $A(-1, 2)$ to the circle given by $x^2+y^2=5$.

419. Set up an equation for the tangent at the point $A(-5, 7)$ to the circle given by $(x+2)^2+(y-3)^2=25$.

420. On the circle given by $16x^2+16y^2+48x-8y-43=0$ find the point M_1 which is closest to the straight line given by $8x-4y+73=0$ and calculate the distance d from M_1 to this line.

421. Set up an equation for the tangent at the point $M_1(x_1, y_1)$ to the circle given by $x^2+y^2=R^2$.

422. Set up an equation for the tangent at the point $M_1(x_1, y_1)$ to the circle given by $(x-\alpha)^2+(y-\beta)^2=R^2$.

423. Find the acute angle formed by the intersection of the straight line and circle given by $3x-y-1=0$ and $(x-2)^2+ +y^2=5$. (By the angle between a straight line and a circle is meant the angle between the line and the tangent to the circle at their point of intersection.)

424. Find the angle at which the two circles given by

$$(x-3)^2+(y-1)^2=8, \quad (x-2)^2+(y+2)^2=2$$

intersect. (By the angle between two circles is meant the angle between their tangents at the point of intersection.)

425. Derive the condition for the two circles given by

$$(x-\alpha_1)^2+(y-\beta_1)^2=R_1^2, \quad (x-\alpha_2)^2+(y-\beta_2)^2=R_2^2$$

to cut at right angles.

426. Show that the two circles given by

$$x^2+y^2-2mx-2ny-m^2+n^2=0$$
$$x^2+y^2-2nx+2my+m^2-n^2=0$$

intersect at right angles.

427. Set up equations for the tangents drawn from the point $A(5/3, -5/3)$ to the circle given by $x^2+y^2=5$.

428. Set up equations for the tangents drawn from the point $A(1, 6)$ to the circle given by $x^2+y^2+2x-19=0$.

429. Find equations for the members of the pencil of straight lines given by $\alpha(3x+4y-10)+\beta(3x-y-5)=0$ which touch the circle given by $x^2+y^2+2x-4y=0$.

430. Find the angle between the tangents drawn from the point $A(4, 2)$ to the circle given by $x^2+y^2=10$.

431. Set up an equation for the chord drawn through the points of contact of the tangents drawn from the point $P(2, -3)$ to the circle given by $(x-1)^2+(y+5)^2=4$.

432. Calculate the distance from the point $C(6, -8)$ to the chord through the points of contact of the tangents from C to the circle given by $x^2+y^2=25$.

433. Tangents are drawn from the point $P(-9, 3)$ to the circle given by $x^2+y^2-6x+4y-78=0$. Calculate the distance d from the centre of the circle to the chord through the points of contact of these tangents.

434. Calculate the length d of the chord joining the points of contact of the tangents drawn from the point $M(4, -4)$ to the circle given by

$$x^2+y^2-6x+2y+5=0.$$

435. Calculate the length of a tangent drawn from the point $A(1, -2)$ to the circle given by $x^2+y^2+x-3y-3=0$.

436. Set up equations for the tangents to the circle given by $x^2+y^2+10x-2y+6=0$ which are parallel to the straight line given by $2x+y-7=0$.

437. Set up equations for the tangents to the circle given by $x^2+y^2-2x+4y=0$ which are perpendicular to the straight line given by $x-2y+9=0$.

438. Set up the polar equation of a circle, given its radius R and the polar coordinates of its centre $C(R, \theta_0)$.

439. Set up the polar equation of a circle, given its radius R and that the polar coordinates of its centre are:

1) $C(R, 0)$ 2) $C(R, \pi)$ 3) $C(R, \pi/2)$ 4) $C(R, -\pi/2)$.

440. Find the radius and the polar coordinates of the centre of each of the circles given by:

1) $\varrho=4 \cos \theta$ 2) $\varrho=3 \sin \theta$ 3) $\varrho=-2 \cos \theta$
4) $\varrho=-5 \sin \theta$ 5) $\varrho=6 \cos (\pi/3-\theta)$
6) $\varrho=8 \sin (\theta-\pi/3)$ 7) $\varrho=8 \sin (\pi/3-\theta)$

441. Find equations in rectangular Cartesian coordinates for the circles given by the following equations, assuming that the pole and origin, and the polar axis and the positive x-axis, coincide.

1) $\varrho=3 \cos \theta$ 2) $\varrho=-4 \sin \theta$ 3) $\varrho=\cos \theta-\sin \theta$.

442. Find the polar equations for the circles given by the following equations, if the pole and origin, and the polar axis and the positive x-axis, coincide.

1) $x^2+y^2=x$ 2) $x^2+y^2=-3x$ 3) $x^2+y^2=5y$
4) $x^2+y^2=-y$ 5) $x^2+y^2=x+y$.

443. Set up the polar equation for the tangent at the point $M_1(R, \theta_0)$ to the circle given by $\varrho=R$.

§ 18. The Ellipse

The locus of a point which moves in a plane so that the sum of its distances from two fixed points in the plane (called the foci) is always a constant quantity (and greater than the distance between the foci) is called an ellipse. The foci are usually denoted by F_1 and F_2, and the distance between them by $2c$; and if M is an arbitrary point of the ellipse, the constant sum F_1M+F_2M is usually denoted by $2a$. It follows from the definition that $2a > 2c$, or $a > c$.

The equation of an ellipse referred to a rectangular Cartesian coordinate system which has its x-axis along the line of foci F_1F_2 and its origin mid-way between F_1 and F_2 has the form

$$\frac{x^2}{a^2}+\frac{y^2}{b^2} = 1 \tag{1}$$

where $b = \sqrt{(a^2-c^2)}$. Clearly, $a > b$.

(1) is called the canonical equation† of an ellipse.

With the chosen coordinate system, the coordinate axes are axes of symmetry of the ellipse, and the origin is a centre of symmetry (Fig. 12). The axes of symmetry of an ellipse are simply called its axes, and its centre of symmetry is simply called its centre. The points where the ellipse cuts the axes are called its vertices. In Fig. 12 the vertices of the ellipse are the points A', A, B', B. The intervals $A'A = 2a$ and $B'B = 2b$ are themselves often called the axes of the ellipse; and the interval $OA = a$ is called the major semi-axis and the interval $OB = b$ the minor semi-axis.

If the foci of the ellipse lie on the y-axis (symmetrically about the origin), the equation of the ellipse still has the same form (1), but now $b > a$; consequently, if we wished the letter a to denote the major semi-axis, it would be necessary to interchange the a and the b in (1). However, for convenience in formulating problems, we shall stipulate that the letter a shall

† i.e. *standard* equation.

always denote the semi-axis situated on the x-axis, and the letter b the semi-axis situated on the y-axis, irrespective of whether a is greater than b or not. If $a = b$, then equation (1) defines a circle, which is therefore a particular case of an ellipse.

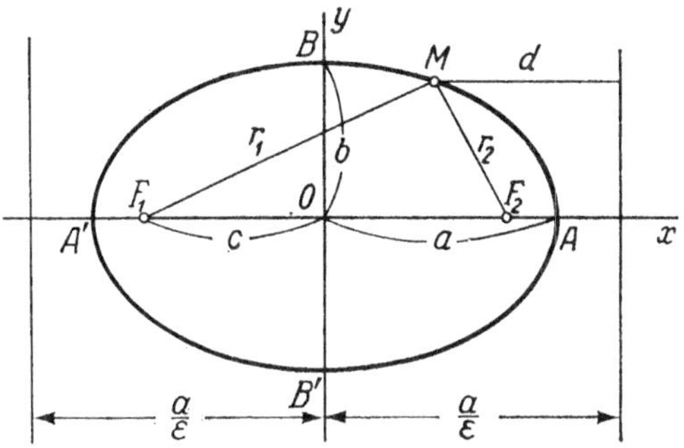

FIG. 12

The number $\varepsilon = \frac{c}{a}$, where a is the major semi-axis, is called the eccentricity of the ellipse. Clearly, $\varepsilon < 1$ (for a circle, $\varepsilon = 0$).

If $M(x, y)$ is an arbitrary point of the ellipse, the intervals

$$F_1M = r_1 \quad \text{and} \quad F_2M = r_2$$

are called the focal radii of the point M. The focal radii may be calculated from the formulae

$$r_1 = a + \varepsilon x, \quad r_2 = a - \varepsilon x.$$

For an ellipse defined by equation (1) with $a > b$, the straight lines given by

$$x = -\frac{a}{\varepsilon}, \quad x = +\frac{a}{\varepsilon}$$

are called the directrices of the ellipse. (If $b > a$, then the directrices are given by $y = -b/\varepsilon$, $y = +b/\varepsilon$.)

Either directrix has this property: if r is the distance of any point of the ellipse from one of the foci, and if d is the distance of this point from the directrix which is nearer to this focus, then the ratio r/d is always equal to the eccentricity ε of the ellipse:

$$r/d = \varepsilon.$$

If two planes α, β intersect at an acute angle φ, then the projection on to the plane β of a circle of radius a lying in the plane α is an ellipse with major semi-axis a and minor semi-axis b given by $b = a \cos \varphi$. See Fig. 13.

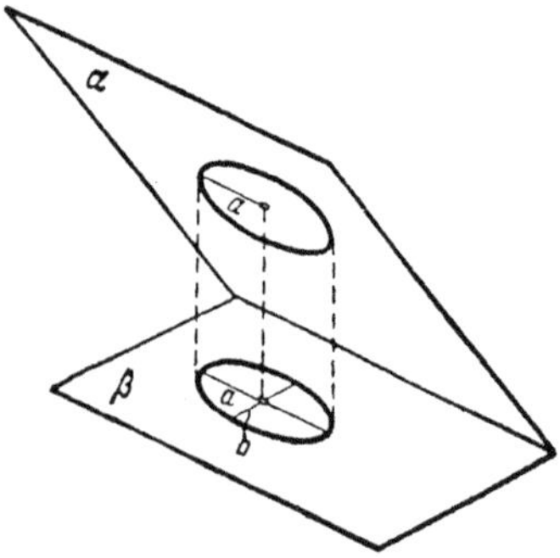

FIG. 13

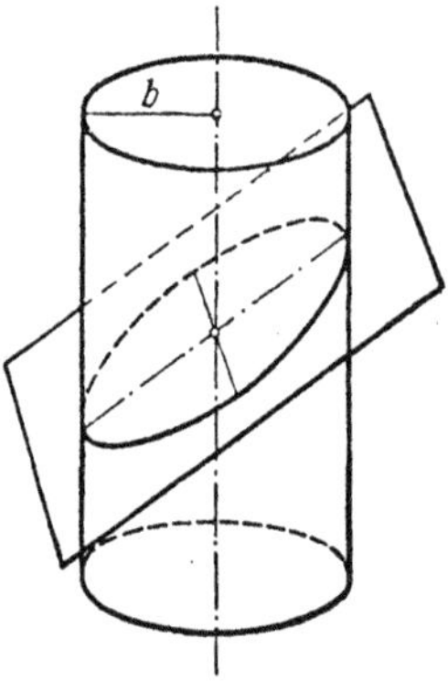

FIG. 14

If the directrix of a circular cylinder is a circle of radius b, then a plane inclined at an acute angle φ to the axis of this cylinder will intersect it in an ellipse having its minor semi-axis equal to b and its major semi-axis a, where $a = b/(\sin \varphi)$. See Fig. 14.

444. Set up an equation for an ellipse whose foci lie on the x-axis and are equidistant from the origin, given that:

1) its semi-axes are equal to 5 and 2

2) its major axis is equal to 10 and the distance between its foci is $2c = 8$

3) its minor axis is equal to 24 and the distance between its foci is $2c = 10$

4) the distance between its foci is $2c = 6$ and its eccentricity $\varepsilon = 3/5$

5) its major axis is equal to 20 and its eccentricity $\varepsilon = 3/5$

6) its minor axis is equal to 10 and its eccentricity $\varepsilon = 12/13$

7) the distance between its directrices is equal to 5 and the distance between its foci is $2c = 4$

8) its major axis is equal to 8 and the distance between its directrices is equal to 16

9) its minor axis is equal to 6 and the distance between its directrices is equal to 13

10) the distance between its directrices is equal to 32 and $\varepsilon = 1/2$.

445. Set up an equation for an ellipse whose foci lie on the y-axis and are equidistant from the origin, given that:

1) its semi-axes are equal to 7 and 2

2) its major axis is equal to 10 and the distance between the foci is $2c = 8$

3) the distance between its foci is $2c = 24$ and the eccentricity $\varepsilon = 12/13$

4) its minor axis is equal to 16 and the eccentricity $\varepsilon = 3/5$

5) the distance between its foci is $2c = 6$ and the distance between its directrices is $50/3$

6) the distance between its directrices is $22/3$ and the eccentricity $\varepsilon = 3/4$.

446. Find the semi-axes of the ellipses given by:

1) $x^2/16 + y^2/9 = 1$ 2) $x^2/4 + y^2 = 1$ 3) $x^2 + 25y^2 = 25$

4) $x^2 + 5y^2 = 15$ 5) $4x^2 + 9y^2 = 25$ 6) $9x^2 + 25y^2 = 1$

7) $x^2 + 4y^2 = 1$ 8) $16x^2 + y^2 = 16$ 9) $25x^2 + 9y^2 = 1$

10) $9x^2 + y^2 = 1$.

447. Find 1) the semi-axes, 2) the foci, 3) the eccentricity, 4) the equations of the directrices, of the ellipse given by

$$9x^2+25y^2=225.$$

448. Calculate the area of a quadrangle two of whose vertices are the foci of the ellipse given by $x^2+5y^2=20$, its other two vertices being at the ends of the minor axis of the ellipse.

449. Find 1) the semi-axes, 2) the foci, 3) the eccentricity, 4) the equations of the directrices, of the ellipse given by

$$9x^2+5y^2=45.$$

450. Calculate the area of a quadrangle which has two vertices at the foci of the ellipse given by $9x^2+5y^2=1$ and the other two vertices at the ends of the minor axis of the ellipse.

451. Calculate the distance from the focus $F(c, 0)$ of the ellipse given by $x^2/a^2+y^2/b^2=1$ to the nearer of its directrices.

452. Given the rectangular coordinate axes and the unit interval, construct, using compasses only, the foci of the ellipse given by

$$x^2/16+y^2/9=1.$$

453. Find the points on the ellipse given by $x^2/25+y^2/4=1$ whose x-coordinates are equal to -3.

454. Which of the points $A_1(-2, 3)$, $A_2(2, -2)$, $A_3(2, -4)$, $A_4(-1, 3)$, $A_5(-4, -3)$, $A_6(3, -1)$, $A_7(3, -2)$, $A_8(2, 1)$, $A_9(0, 15)$, $A_{10}(0, -16)$ lie on the ellipse given by $8x^2+5y^2=77$? Which points lie inside it and which outside?

455. What sort of curves are defined by the following equations? Draw their graphs.

$$1)\ y = +\frac{3}{4}\sqrt{(16-x^2)} \quad 2)\ y = -\frac{5}{3}\sqrt{(9-x^2)}$$

$$3)\ x = -\frac{2}{3}\sqrt{(9-y^2)} \quad 4)\ x = +\frac{1}{7}\sqrt{(49-y^2)}.$$

456. The eccentricity of an ellipse is $\varepsilon = 2/3$ and the focal radius of a point M of the ellipse is equal to 10. Calculate the distance of the point M from the directrix which is the nearer to this focus.

457. The eccentricity of an ellipse is $\varepsilon = 2/5$ and the distance of a point M of the ellipse from one directrix is equal to 20. Calculate the distance of the point M from the focus which is the nearer to this directrix.

458. The point $M_1(2, -5/3)$ lies on the ellipse given by $x^2/9+y^2/5 = 1$. Set up equations for the straight lines on which the focal radii of the point M lie.

459. Verify that the point $M_1(-4, 2{\cdot}4)$ lies on the ellipse given by $x^2/25+y^2/16 = 1$ and calculate the focal radii of M_1.

460. The eccentricity of an ellipse is $\varepsilon = 1/3$, the centre is at the origin, and $F(-2, 0)$ is one of the foci. Calculate the distance of a point M_1 of the ellipse from the directrix nearer to the given focus if the x-coordinate of M_1 is equal to 2.

461. The eccentricity of an ellipse is $\varepsilon = 1/2$, the centre is at the origin, and one of the directrices is given by $x = 16$. Calculate the distance of a point M_1 of the ellipse from the focus nearer to the given directrix if the x-coordinate of the point M_1 is equal to -4.

462. Find the points of the ellipse given by $x^2/100+y^2/36 = 1$ whose distance from the right-hand focus is equal to 14.

463. Find the points of the ellipse given by $x^2/16+y^2/7 = 1$ whose distance from the left-hand focus is equal to 2·5.

464. A line is drawn through a focus of the ellipse given by $x^2/25+y^2/15 = 1$ perpendicular to its major axis. Find the distances of the points of intersection of this line with the ellipse from the foci.

465. Set up an equation for an ellipse whose foci are on the x-axis and equidistant from the origin, given that:

1) the point $M_1(-2\sqrt{5}, 2)$ is on the ellipse and its minor semi-axis $b = 3$

2) the point $M_1(2, -2)$ is on the ellipse and its major semi-axis $a = 4$

3) the points $M_1(4, -\sqrt{3})$ and $M_2(2\sqrt{2}, 3)$ are on the ellipse

4) the point $M_1(\sqrt{15}, -1)$ is on the ellipse and the distance between its foci is $2c = 8$

5) the point $M_1(2, -5/3)$ is on the ellipse and its eccentricity $\varepsilon = 2/3$

6) the point $M_1(8, 12)$ is on the ellipse and its distance from the left-hand focus is $r_1 = 20$

7) the point $M_1(-\sqrt{5}, 2)$ is on the ellipse and the distance between its directrices is 10.

466. Find the eccentricity ε of an ellipse if:

1) its minor axis subtends an angle 60° at either focus

2) the interval between the foci subtends a right angle at the vertex of the minor axis

3) the distance between directrices is three times the distance between the foci

4) the vertex of the ellipse is the mid-point of the perpendicular dropped from the centre on to a directrix.

467. A line drawn through a focus F perpendicular to the major axis cuts the ellipse in C and the line OC is parallel to the line through the vertices A, B of the ellipse (Fig. 15). What is the eccentricity ε of this ellipse?

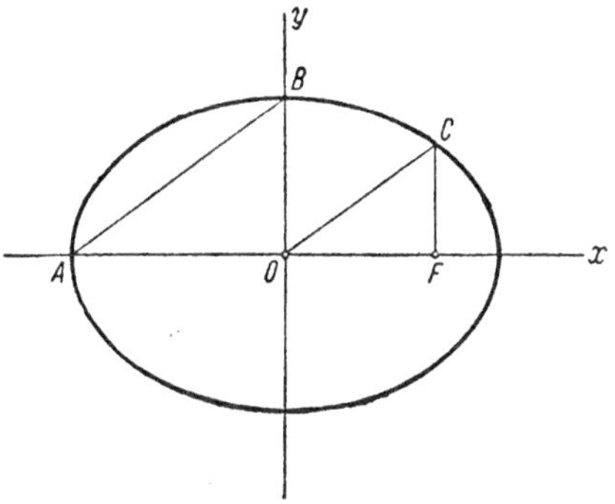

FIG. 15

468. Set up an equation for an ellipse with semi-axes a, b and centre $C(x_0, y_0)$ if its axes of symmetry are parallel to the coordinate axes.

469. Set up an equation for an ellipse which touches the x-axis at the point $A(3, 0)$ and the y-axis at the point $B(0, -4)$ if its axes of symmetry are parallel to the coordinate axes.

470. Set up an equation for the ellipse which has its centre at the point $C(-3, 2)$, touches both coordinate axes, and has its axes of symmetry parallel to them.

471. Show that each of the following equations defines an ellipse and find the coordinates of the centre, the semi-axes, eccentricity, and the equations of the directrices:

1) $5x^2+9y^2-30x+18y+9=0$

2) $16x^2+25y^2+32x-100y-284=0$

3) $4x^2+3y^2-8x+12y-32=0.$

472. What sort of curves are defined by the following equations? Draw their graphs.

1) $y = -7+\frac{2}{5}\sqrt{(16+6x-x^2)}$ 2) $y = 1-\frac{4}{3}\sqrt{(-6x-x^2)}$

3) $x = -2\sqrt{(-5-6y-y^2)}$ 4) $x = -5+\frac{2}{3}\sqrt{(8+2y-y^2)}$

473. Set up an equation for an ellipse given that:

1) its major axis is equal to 26 and the foci are $F_1(-10, 0)$, $F_2(14, 0)$

2) its minor axis is equal to 2 and the foci are $F_1(-1, -1)$, $F_2(1, 1)$

3) its foci are $F_1(-2, 3/2)$, $F_2(2, -3/2)$, and its eccentricity $\varepsilon = \sqrt{2}/2$

4) its foci are $F_1(1, 3)$, $F_2(3, 1)$ and the distance between the directrices is equal to $12\sqrt{2}$.

474. Set up an equation for an ellipse which has eccentricity $\varepsilon = 2/3$, a focus at $F(2, 1)$, and the directrix corresponding to this focus given by $x-5 = 0$.

475. Set up an equation for an ellipse which has eccentricity $\varepsilon = 1/2$, a focus at $F(-4, 1)$, and the directrix corresponding to this focus given by $y+3 = 0$.

476. Set up an equation for an ellipse which goes through the point $A(-3, -5)$, has a focus at $F(-1, -4)$, and the directrix corresponding to this focus given by $x-2 = 0$.

477. Set up an equation for an ellipse which has eccentricity $\varepsilon = 1/2$, a focus at $F(3, 0)$, and the directrix corresponding to this focus given by $x+y-1 = 0$.

478. A focus and corresponding directrix of an ellipse are given by $F(1, 0)$ and $2x-y-10=0$. Set up an equation for this ellipse if it goes through the point $M_1(2, -1)$.

479. The point $M_1(3, -1)$ is the end of the minor axis of an ellipse whose foci lie on the line given by $y+6=0$. Set up an equation for this ellipse if its eccentricity is $\varepsilon = \sqrt{2}/2$.

480. Find the points of intersection of the straight line and ellipse given by $x+2y-7=0$ and $x^2+4y^2=25$.

481. Find the points of intersection of the straight line and ellipse given by $3x+10y-25=0$ and $x^2/25+y^2/4=1$.

482. Find the points of intersection of the straight line and ellipse given by $3x-4y-40=0$ and $x^2/16+y^2/9=1$.

483. How are the straight line and ellipse given by the following equations situated? Does the straight line intersect, touch, or pass outside the ellipse?

1) $2x-y-3=0, \quad x^2/16+y^2/9=1$
2) $2x+y-10=0, \quad x^2/9+y^2/4=1$
3) $3x+2y-20=0, \quad x^2/40+y^2/10=1.$

484. For what values of c will the straight line given by $y=-x+c$, 1) intersect the ellipse given by $x^2/20+y^2/5=1$, 2) touch it, 3) pass outside it?

485. Derive the condition for the straight line given by $y=mx+c$ to touch the ellipse given by $x^2/a^2+y^2/b^2=1$.

486. Set up an equation for the tangent at the point $M_1(x_1, y_1)$ on the ellipse given by $x^2/a^2+y^2/b^2=1$.

487. Show that the tangents to an ellipse, given by $x^2/a^2 + y^2/b^2 = 1$, at the ends of any diameter are parallel. (A chord of an ellipse which passes through its centre is called a diameter.)

488. Set up equations for the tangents to the ellipse given by $x^2/10+2y^2/5 = 1$ which are parallel to the straight line given by $3x+2y+7 = 0$.

489. Set up equations for the tangents to the ellipse given by $x^2+4y^2 = 20$ which are perpendicular to the straight line given by $2x-2y-13 = 0$.

490. Find equations for the tangents to the ellipse given by $x^2/30+y^2/24 = 1$ which are parallel to the straight line given by $4x-2y+23 = 0$ and find the distance between these tangents.

491. On the ellipse given by $x^2/18+y^2/8 = 1$ find the point M_1 which is closest to the line given by $2x-3y+25 = 0$ and calculate the distance d from the point M_1 to this line.

492. Find equations for the tangents drawn from the point $A(10/3, 5/3)$ to the ellipse given by $x^2/20+y^2/5 = 1$.

493. Find an equation for the chord joining the points of contact of the tangents drawn from the point $C(10, -8)$ to the ellipse given by $x^2/25+y^2/16 = 1$.

494. Tangents are drawn from the point $P(-16, 9)$ to the ellipse given by $x^2/4+y^2/3 = 1$. Calculate the distance d from the point P to the chord of the ellipse joining the points of contact of these tangents.

495. Find an equation for the ellipse which goes through the point $A(4, -1)$ and touches the straight line given by $x+4y-10=0$ and whose axes coincide with the coordinate axes.

496. Find an equation for the ellipse which touches the straight lines given by $3x-2y-20=0$ and $x+6y-20=0$ and whose axes coincide with the coordinate axes.

497. O is the centre of an ellipse. The tangent at any point M of the ellipse cuts the focal axis in T, and N is the foot of the perpendicular from M on to the focal axis. Prove that $OT.ON$ is always equal to a^2, where a is the major semi-axis of the ellipse.

498. P_1, P_2 are the feet of the perpendicular from the foci F_1, F_2 on to any tangent to an ellipse. Prove that $F_1P_1.F_2P_2$ is always equal to b^2, where b is the minor semi-axis of the ellipse.

499. The straight line given by $x-y-5=0$ touches an ellipse whose foci are $F_1(-3, 0)$, $F_2(3, 0)$. Find an equation for this ellipse.

500. Find an equation for the ellipse whose foci are on the x-axis and equidistant from the origin and whose minor axis is $b=2$, if it is also known that the straight line given by $3x+10y-25=0$ touches the ellipse.

501. Prove that the tangent at any point M of an ellipse is equally inclined to the focal radii F_1M, F_2M and that it is the external bisector of the angle F_1MF_2.

502. A ray of light is projected from the left-hand focus of the ellipse given by $x^2/45+y^2/20=1$ so that it makes an obtuse angle α with the axis Ox, and it is known that $\tan \alpha = -2$. The ray is reflected on striking the ellipse. Find an equation for the straight line on which the reflected ray lies.

503. Find the points of intersection of the ellipses given by

$$x^2+9y^2-45 = 0, \quad x^2+9y^2-6x-27 = 0.$$

504. Verify that the ellipses given by

$$n^2x^2+m^2y^2-m^2n^2 = 0, \quad m^2x^2+n^2y^2-m^2n^2 = 0, \quad (m \neq n)$$

intersect in four points which lie on a circle with centre at the origin and find the radius R of this circle.

505. Two planes α, β intersect at an angle $\varphi = 30°$. Find the semi-axes of the ellipse which is the projection on plane β of a circle of radius $R = 10$ lying in the plane α.

506. An ellipse with minor semi-axis $b = 6$ is the projection of a circle of radius $R = 12$. Find the angle φ between the planes of the circle and the ellipse.

507. The directrix of a circular cylinder is a circle of radius $R = 8$. Find the semi-axes of the ellipse which is obtained as the curve of intersection of this cylinder with a plane inclined at an angle $\varphi = 30°$ to its axis.

508. The directrix of a circular cylinder is a circle of radius $R = \sqrt{3}$. At what angle must a plane be inclined to the axis of the cylinder so that its intersection may be an ellipse with major semi-axis $a = 2$?

509. A transformation of points in the plane under which $M(x, y)$ goes over into $M'(x', y')$ and

$x' = x$

$y' = qy$, where q is a positive constant,

is called a uniform compression in the direction Oy if $q<1$ (or a uniform extension if $q>1$). See Fig. 16.

Similarly, a uniform compression (or extension) in the direction Ox is given by

$$x' = qx$$
$$y' = y. \quad \text{(See Fig. 17.)}$$

Into what sort of curve will the circle given by $x^2+y^2 = 25$ transform under a uniform compression in the direction Oy with coefficient $q = 4/5$?

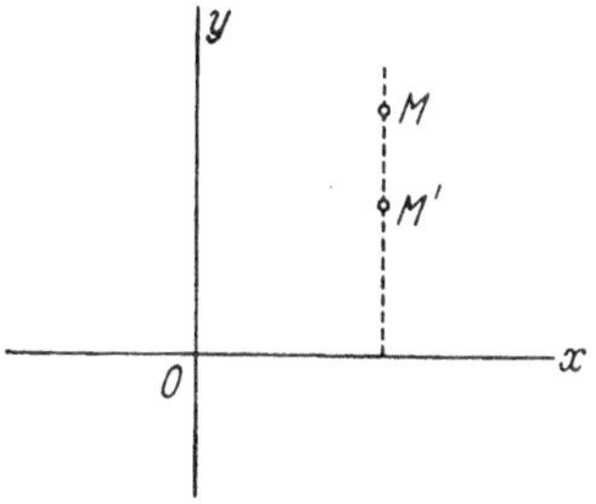

FIG. 16

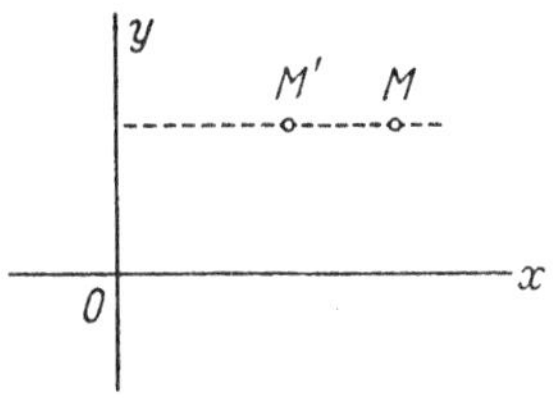

FIG. 17

510. The coefficient of a uniform compression of the plane in the direction Ox is 3/4. Find an equation for the curve into which the ellipse given by $x^2/16+y^2/9 = 1$ is transformed under this compression.

511. Find an equation for the curve into which the ellipse given by $x^2/49+y^2/9 = 1$ is transformed under two consecutive uniform compressions of the plane, the first in the direction Oy with coefficient 4/3, and the second in the direction Ox with coefficient 6/7.

512. Find the coefficient of the uniform compression of the plane in the direction Oy under which the ellipse given by $x^2/36+y^2/9 = 1$ transforms into another ellipse given by $x^2/36+y^2/16 = 1$.

513. Find the coefficient of the uniform compression of the plane in the direction Ox under which the ellipse given by $x^2/81+y^2/25 = 1$ transforms into another ellipse given by $x^2/36+y^2/25 = 1$.

514. Find the coefficients q_1, q_2 of two successive compressions of the plane in the directions Oy, Ox respectively, under which the ellipse given by $x^2/25+y^2/9 = 1$ is transformed into the circle given by $x^2+y^2 = 16$.

§ 19. The Hyperbola

The locus of a point which moves in a plane so that the absolute value of the difference of its distances from two fixed points of the plane (called the foci) is always a constant quantity (and less than the distance between the foci) is called a hyperbola. The foci are usually denoted by F_1 and F_2, and the distance between them by $2c$; and if M is an arbitrary point of the hyperbola, the constant value of $|F_1M-F_2M|$ is usually denoted by $2a$. It follows from the definition that $2a<2c$, or $a<c$.

The equation of a hyperbola referred to a rectangular Cartesian coordinate system which has its x-axis along the line of foci F_1F_2 and its origin mid-way between F_1 and F_2 has the form

$$\frac{x^2}{a^2}-\frac{y^2}{b^2} = 1 \tag{1}$$

where $b=\sqrt{(c^2-a^2)}$. (1) is called the canonical equation of the hyperbola.

With the chosen coordinate system, the coordinate axes are axes of symmetry of the hyperbola, and the origin is a centre of symmetry (Fig. 18). The axes of symmetry of a hyperbola are simply called its axes, and its centre of symmetry is simply called its centre. The hyperbola cuts one of its axes, in two points which are called its vertices; these are the points A' and A in Fig. 18.

The rectangle with sides $2a$ and $2b$, placed symmetrically with respect to the axes and touching the vertices of the hyperbola, is called the main rectangle of the hyperbola.

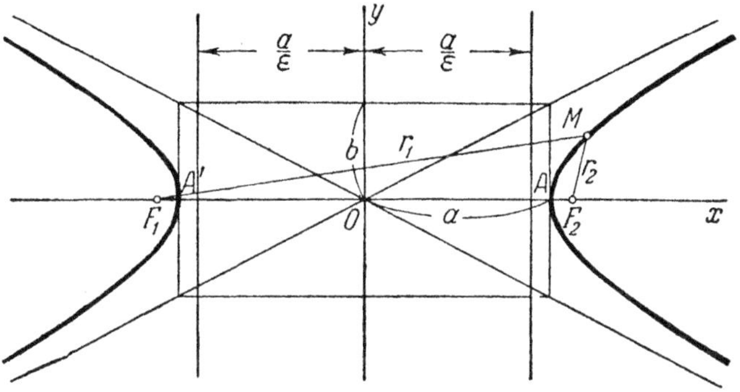

FIG. 18

The intervals of length $2a$ and $2b$ joining the mid-points of the sides of the main rectangle also are referred to as the axes of the hyperbola. The lines on which the diagonals of the main rectangle lie are called the asymptotes of the hyperbola; their equations are

$$y=\frac{b}{a}x \quad \text{and} \quad y=-\frac{b}{a}x.$$

The equation
$$-\frac{x^2}{a^2}+\frac{y^2}{b^2}=1 \qquad (2)$$

defines a hyperbola which is symmetrically situated with respect to the coordinate axes and which has its foci on the y-axis. Equation (2), like (1), is called a canonical equation† of a hyperbola. In this case, the constant difference of distances of any point on the hyperbola from the foci is equal to $2b$.

† i.e. *standard* equation.

The two hyperbolae whose equations in the same system of coordinates are

$$\frac{x^2}{a^2}-\frac{y^2}{b^2}=1 \quad \text{and} \quad -\frac{x^2}{a^2}+\frac{y^2}{b^2}=1$$

are called *conjugate hyperbolae.*

A hyperbola with equal semi-axes ($a = b$) is called an equilateral or rectangular hyperbola; its canonical equation has the form

$$x^2-y^2=a^2 \quad \text{or} \quad -x^2+y^2=a^2 .$$

The number
$$\varepsilon = c/a$$

where a is the distance from the centre of the hyperbola to its vertex, is called the eccentricity of the hyperbola. For any hyperbola, $\varepsilon > 1$. If M is any point on a hyperbola, the intervals F_1M, F_2M are called the focal radii of the point M. The focal radii of a point on the right branch of a hyperbola are given by

$$r_1 = \varepsilon x+a, \quad r_2 = \varepsilon x-a$$

and the focal radii of a point on the left branch of the hyperbola are given by

$$r_1 = -\varepsilon x-a, \quad r_2 = -\varepsilon x+a.$$

If a hyperbola is defined by equation (1), the straight lines given by $x = -a/\varepsilon$, $x = a/\varepsilon$ are called the directrices of the hyperbola. For the hyperbola given by equation (2), the directrices are given by $y = -b/\varepsilon$, $y=b/\varepsilon$.

A directrix has this property:

if r is the distance of any point on the hyperbola from a focus, and

if d is the distance of that point from the directrix corresponding to this focus (i.e., the directrix nearer to the focus),

then the ratio r/d is always equal to the eccentricity ε: $r/d = \varepsilon$.

515. Set up an equation for a hyperbola whose foci lie on the x-axis and are equidistant from the origin, given that:

1) its axes are $2a = 10$ and $2b = 8$

2) the distance between the foci is $2c = 10$ and the minor axis is $2b = 8$

3) the distance between the foci is $2c = 6$ and the eccentricity is $\varepsilon = 3/2$

4) the major axis is $2a = 16$ and the eccentricity $\varepsilon = 5/4$

5) the distance between the foci is $2c = 20$ and the equations of the asymptotes are $y = \pm 4x/3$

6) the distance between the directrices is $288/13$ and the distance between the foci is $2c = 26$

7) the distance between the directrices is $32/5$ and the minor axis $2b = 6$

8) the distance between the directrices is $8/3$ and the eccentricity is $\varepsilon = 3/2$

9) the distance between the directrices is $64/5$ and the equations of the asymptotes are $y = \pm 3x/4$.

516. Set up an equation for a hyperbola whose foci lie on the y-axis and are equidistant from the origin, given that:

1) its semi-axes are $a = 6$ (along the x-axis) and $b = 18$

2) the distance between the foci is $2c = 10$ and the eccentricity $\varepsilon = 5/3$

3) the distance between vertices is 48 and the equations of the asymptotes are $y = \pm 12x/5$

4) the distance between the directrices is $50/7$ and the eccentricity is $\varepsilon = 7/5$

5) the distance between the directrices is $32/5$ and the equations of the asymptotes are $y = \pm 4x/3$.

517. Find the semi-axes a and b of hyperbolae given by:

1) $x^2/9 - y^2/4 = 1$ 2) $x^2/16 - y^2 = 1$ 3) $x^2 - 4y^2 = 16$

4) $x^2 - y^2 = 1$ 5) $4x^2 - 9y^2 = 25$ 6) $25x^2 - 16y^2 = 1$

7) $9x^2 - 64y^2 = 1$.

518. For the hyperbola given by $16x^2-9y^2 = 144$ find 1) the semi-axes a and b, 2) the foci, 3) the eccentricity, 4) the equations of the asymptotes, 5) the equations of the directrices.

519. For the hyperbola given by $16x^2-9y^2 = -144$ find 1) the semiaxes a and b, 2) the foci, 3) the eccentricity, 4) the equations of the asymptotes, 5) the equations of the directrices.

520. Calculate the area of the triangle formed by the asymptotes of the hyperbola given by $x^2/4-y^2/9 = 1$ with the straight line defined by $9x+2y-24 = 0$.

521. What sort of curves are defined by the following equations? Draw their graphs.

$$1)\ y = +\frac{2}{3}\sqrt{(x^2-9)} \quad 2)\ y = -3\sqrt{(x^2+1)}$$

$$3)\ x = -\frac{4}{3}\sqrt{(y^2+9)} \quad 4)\ y = +\frac{2}{5}\sqrt{(x^2+25)}.$$

522. The point $M_1(10, -\sqrt{5})$ lies on the hyperbola given by $x^2/80-y^2/20 = 1$. Find equations for the straight lines on which the focal radii of M_1 lie.

523. Verify that the point $M_1(-5, 9/4)$ lies on the hyperbola given by $x^2/16-y^2/9 = 1$ and find the focal radii of M_1.

524. The eccentricity of a hyperbola is $\varepsilon = 2$, and one focal radius of a point M on the hyperbola is equal to 16. Calculate the distance of the point M from the directrix corresponding to this focus.

525. A point M on a hyperbola of eccentricity $\varepsilon = 3$ is at a distance equal to 4 from one directrix. Calculate the distance of the point M from the focus corresponding to this directrix.

526. A hyperbola of eccentricity $\varepsilon = 2$ has its centre at the origin and one focus at $F(12, 0)$. Calculate the distance from the directrix corresponding to the given focus of a point M_1 on the hyperbola, if the x-coordinate of M_1 is equal to 13.

527. A hyperbola has eccentricity $\varepsilon = 3/2$, its centre is at the origin, and one of its directrices is given by $x = -8$. Calculate the distance of a point M_1 on the hyperbola from the focus corresponding to the given directrix, if the x-coordinate of M_1 is equal to 10.

528. Find the points on the hyperbola given by $x^2/64 - y^2/36 = 1$ whose distances from the right-hand focus are equal to 4·5.

529. Find the points on the hyperbola given by $x^2/9 - y^2/16 = 1$ whose distances from the left-hand focus are equal to 7.

530. A line is drawn through the left-hand focus of the hyperbola given by $x^2/144 - y^2/25 = 1$ perpendicular to the focal axis. Find the focal distances of the points of intersection of this line with the hyperbola.

531. Given the rectangular coordinate axes and the unit-interval, construct, using compasses only, the foci of the hyperbola given by $x^2/16 - y^2/25 = 1$.

532. Find an equation for a hyperbola whose foci lie on the x-axis and are equidistant from the origin, given that:

1) the points $M_1(6, -1)$, $M_2(-8, 2\sqrt{2})$ lie on the hyperbola

2) the point $M_1(-5, 3)$ lies on the hyperbola and its eccentricity $\varepsilon = \sqrt{2}$

3) the point $M_1(9/2, -1)$ lies on the hyperbola and the equations of its asymptotes are $y = \pm 2x/3$

4) the point $M_1(-3, 5/2)$ lies on the hyperbola and the equations of its directrices are $x = \pm 4/3$

5) the equations of the asymptotes are $y = \pm 3x/4$ and the equations of the directrices are $x = \pm 16/5$.

533. Find the eccentricity of a rectangular hyperbola.

534. Find the eccentricity of a hyperbola if the interval between its vertices subtends an angle of 60° at a focus of the conjugate hyperbola.

535. Find an equation for a hyperbola of eccentricity $\varepsilon = 2$ whose foci coincide with the foci of the ellipse given by $x^2/25 + y^2/9 = 1$.

536. Find an equation for the hyperbola whose foci lie at the vertices of the ellipse given by $x^2/100 + y^2/64 = 1$ and whose directrices go through the foci of this ellipse.

537. Show that the distance from a focus of the hyperbola given by $x^2/a^2 - y^2/b^2 = 1$ to its asymptotes is equal to b.

538. Show that the product of the distances of any point on the hyperbola given by $x^2/a^2 - y^2/b^2 = 1$ to its asymptotes is constant and equal to $a^2b^2/(a^2+b^2)$.

539. Show that the area of a parallelogram which is bounded by the asymptotes of the hyperbola given by $x^2/a^2 - y^2/b^2 = 1$ and by lines drawn through any point of the hyperbola parallel to the asymptotes is constant and equal to $ab/2$.

540. Set up an equation for a hyperbola which has semi-axes a and b, its centre at $C(x_0, y_0)$, and its foci on a line 1) parallel to Ox, 2) parallel to Oy.

541. Show that each of the following equations defines a hyperbola and find the coordinates of its centre, the semi-axes, eccentricity, and the equations of its asymptotes and of its directrices:

1) $16x^2-9y^2-64x-54y-161 = 0$

2) $9x^2-16y^2+90x+32y-367 = 0$

3) $16x^2-9y^2-64x-18y+199 = 0$

542. What sort of curves are defined by the following equations? Draw their graphs:

1) $y = -1+\frac{2}{3}\sqrt{(x^2-4x-5)}$ 2) $y = 7-\frac{3}{2}\sqrt{(x^2-6x+13)}$

3) $x = 9-2\sqrt{(y^2+4y+8)}$ 4) $x = 5-\frac{3}{4}\sqrt{(y^2+4y-12)}$.

543. Find an equation for a hyperbola given that

1) the distance between its vertices is equal to 24 and the foci are $F_1(-10, 2)$, $F_2(16, 2)$

2) the foci are $F_1(3, 4)$, $F_2(-3, -4)$ and the distance between directrices is equal to 3·6

3) the asymptotes are perpendicular and the foci are $F_1(4, -4)$, $F_2(-2, 2)$.

544. Find an equation for a hyperbola of eccentricity $\varepsilon = 5/4$ which has a focus at $F(5, 0)$ if the corresponding directrix is given by $5x-16 = 0$.

545. Find an equation for a hyperbola of eccentricity $\varepsilon = 13/12$ which has a focus at $F(0, 13)$ if the corresponding directrix is given by $13y-144 = 0$.

546. The point $A(-3, -5)$ lies on a hyperbola which has a focus at $F(-2, -3)$, and the directrix corresponding to this focus is given by $x+1 = 0$. Set up an equation for this hyperbola.

547. Set up an equation for a hyperbola of eccentricity $\varepsilon=\sqrt{5}$ which has a focus at $F(2, -3)$ if the directrix corresponding to this focus is given by $3x-y+3 = 0$.

548. Set up an equation for a hyperbola which goes through the point $M_1(1, -2)$ and has a focus at $F(-2, 2)$ if the corresponding directrix is given by $2x-y-1 = 0$.

549. Find an equation for the rectangular hyperbola given by $x^2-y^2 = a^2$ in a new coordinate system having the axes along the asymptotes of the hyperbola.

550. Verify that each of the following equations defines a hyperbola. Find the centre, semi-axes, and the equations of the asymptotes of each and draw its graph:

1) $xy = 18$ 2) $2xy-9 = 0$ 3) $2xy+25 = 0$.

551. Find the points of intersection of the straight line and hyperbola given by $2x-y-10 = 0$ and $x^2/20-y^2/5 = 1$.

552. Find the points of intersection of the straight line and hyperbola given by $4x-3y-16 = 0$ and $x^2/25-y^2/16 = 1$.

553. Find the points of intersection of the straight line and hyperbola given by $2x-y+1 = 0$ and $x^2/9-y^2/4 = 1$.

554. How are the straight line and hyperbola given by the following equations situated? Does the straight line intersect, touch, or pass outside, the hyperbola?

1) $x-y-3 = 0, \quad x^2/12-y^2/3 = 1$

2) $x-2y+1 = 0, \quad x^2/16-y^2/9 = 1$

3) $7x-5y = 0, \quad x^2/25-y^2/16 = 1.$

555. For what values of c will the straight line given by $y = 5x/2+c$, 1) intersect the hyperbola given by $x^2/9-y^2/36 = 1$ 2) touch it, 3) pass outside it?

556. Derive the condition for the straight line given by $y = mx+c$ to touch the hyperbola given by $x^2/a^2-y^2/b^2 = 1$.

557. Set up an equation for the tangent at the point $M_1(x_1, y_1)$ on the hyperbola given by $x^2/a^2-y^2/b^2 = 1$.

558. Show that tangents to a hyperbola drawn at the ends of any diameter of the curve are parallel.

559. Find equations for the tangents to the hyperbola given by $x^2/20-y^2/5 = 1$ which are perpendicular to the straight line defined by $4x+3y-7 = 0$.

560. Find equations for the tangents to the hyperbola given by $x^2/16-y^2/64 = 1$ which are parallel to the straight line defined by $10x-3y+9 = 0$.

561. Set up equations for the tangents to the hyperbola given by $x^2/16-y^2/8 = -1$ which are parallel to the straight line given by $2x+4y-5 = 0$ and find the distance d between them.

562. On the hyperbola given by $x^2/24-y^2/18 = 1$ find the point M_1 which is closest to the straight line given by $3x+2y+1 = 0$ and calculate the distance from this point to the straight line.

563. Set up equations for the tangents from the point $A(-1, -7)$ to the hyperbola given by $x^2-y^2 = 16$.

564. Find an equation for the chord through the points of contact of the tangents from $C(1, -10)$ to the hyperbola given by $x^2/8-y^2/32 = 1$.

565. Calculate the distance from the point $P(1, -5)$ to the chord through the points of contact of the tangents from P to the hyperbola given by $x^2/3-y^2/5 = 1$.

566. Set up an equation for a hyperbola whose axes coincide with the coordinate axes and which goes through the point $(A\sqrt{6}, 3)$ and touches the straight line given by $9x+2y-$ $-15 = 0$.

567. Set up an equation for a hyperbola whose axes coincide with the coordinate axes and which touches the straight lines defined by

$$5x-6y-16 = 0, \quad 13x-10y-48 = 0.$$

568. Verify that the points of intersection of the hyperbola and ellipse given by $x^2/12-y^2/3 = 1$ and $x^2/20+y^2/5 = 1$ are the vertices of a rectangle, and set up equations for its sides.

569. The tangent at the point M on the hyperbola given by $x^2a^2-y^2/b^2 = 1$ cuts the x-axis in P, and Q is the projection of M on to the x-axis. Show that $OP.OQ = a^2$.

570. Prove that the foci of a hyperbola lie on opposite sides of any tangent to the curve.

571. Show that the product of the distances of the foci from any tangent to the hyperbola given by $x^2/a^2-y^2/b^2 = 1$ is always equal to b^2.

572. The straight line given by $2x-y-4 = 0$ touches a hyperbola having its foci at $F_1(-3, 0)$, $F_2(3, 0)$. Find an equation for this hyperbola.

573. Find an equation for a hyperbola whose foci lie on the x-axis and are equidistant from the origin, given that the distance between vertices is $2a = 8$ and that the straight line given by $15x+16y-36 = 0$ touches the hyperbola.

574. Show that the tangent to a hyperbola at any point M bisects internally the angle between the focal radii F_1M, F_2M.

575. A ray of light is projected from the right-hand focus of the hyperbola given by $x^2/5-y^2/4 = 1$ so that it makes an angle α with the x-axis ($\pi < \alpha < 3\pi/2$). It is known that $\tan \alpha = 2$. The ray is reflected on striking the hyperbola. Find an equation for the straight line on which the reflected ray lies.

576. Prove that any ellipse and hyperbola which have common foci intersect at right angles.

577. The coefficient of a uniform compression of the plane in the direction Oy is 4/3 (see *q.* **509**). Find an equation for the curve into which the hyperbola given by $x^2/16-y^2/9 = 1$ is transformed under this compression.

578. The coefficient of a uniform compression of the plane in the direction Ox is 4/5. Find an equation for the curve into which the hyperbola given by $x^2/25-y^2/9 = 1$ is transformed under this compression.

579. Find an equation for the curve into which the hyperbola given by $x^2-y^2 = 9$ is transformed under two successive, uniform compressions of the plane, the first in the direction Oy with coefficient 2/3, and the second in the direction Ox with coefficient 5/3.

580. Find the coefficient of a uniform compression of the plane in the direction Oy under which the hyperbola given by $x^2/25-y^2/36 = 1$ transforms into another hyperbola given by $x^2/25-y^2/16 = 1$.

581. Find the coefficient of a uniform compression of the plane in the direction Ox under which the hyperbola given by $x^2/4-y^2/9 = 1$ transforms into another hyperbola given by $x^2/16-y^2/9 = 1$.

582. Find the coefficients q_1 and q_2 of the two uniform compressions of the plane in the directions Oy and Ox respectively under which the hyperbola given by $x^2/49-y^2/16 = 1$ is transformed into the hyperbola given by $x^2/25-y^2/64 = 1$.

§ 20. The Parabola

A parabola is the locus of a point which moves in a plane so that its distance from a fixed point of the plane, called the focus, is equal to its distance from a fixed straight line in the plane, called the directrix. The focus is usually denoted by the letter F and its distance from the directrix by $2a$. $2a$ is called the parameter of the parabola

The canonical equation of a parabola is

$$y^2 = 4ax \qquad (1)$$

when it is referred to the rectangular coordinate system shown in Fig. 19, i.e., a system such that the x-axis is along the straight line through the focus perpendicular to the directrix (the direction from the directrix to the focus being regarded as positive), and the origin is midway between directrix and focus.

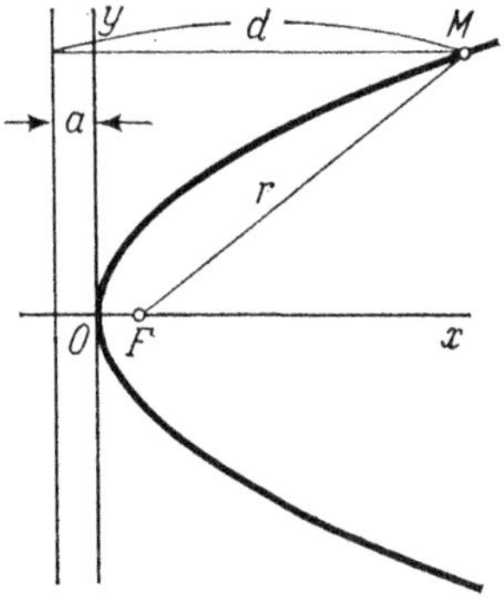

FIG. 19

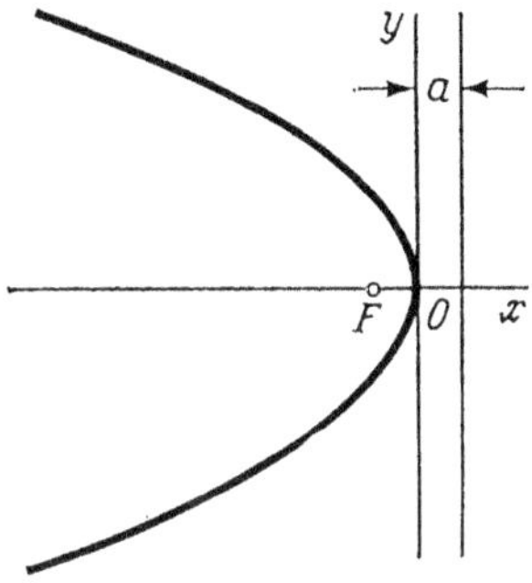

FIG. 20

In this coordinate system the equation of the directrix is $x = -a$.

A parabola has one axis of symmetry which it cuts in just one point. The axis of symmetry is called simply the axis of the parabola, and the point of intersection of the parabola with its axis is called the vertex. For the coordinate system described above, the axis of the parabola coincides with the x-axis, its vertex is at the origin, and the whole parabola lies in the right half-plane.

The equation

$$y^2 = -4ax \tag{2}$$

is the canonical equation for a parabola which has its vertex at the origin, its axis on the x-axis, and which lies entirely in the left half-plane, as shown in Fig. 20.

The equation

$$x^2 = 4ay \tag{3}$$

is the canonical equation for a parabola which has its vertex at the origin, its axis the y-axis, and which lies entirely in the upper half-plane, as shown in Fig. 21.

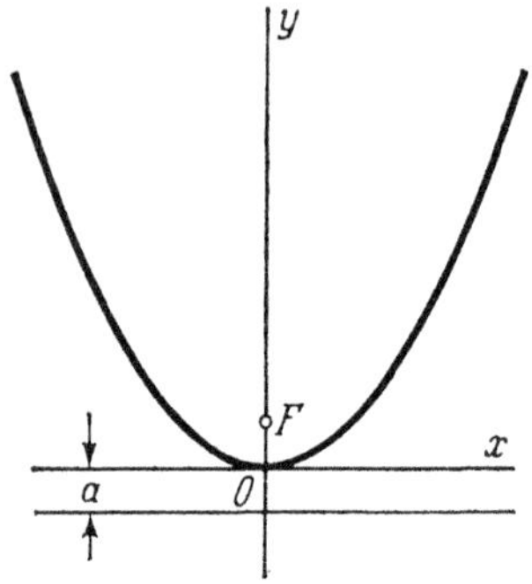

FIG. 21

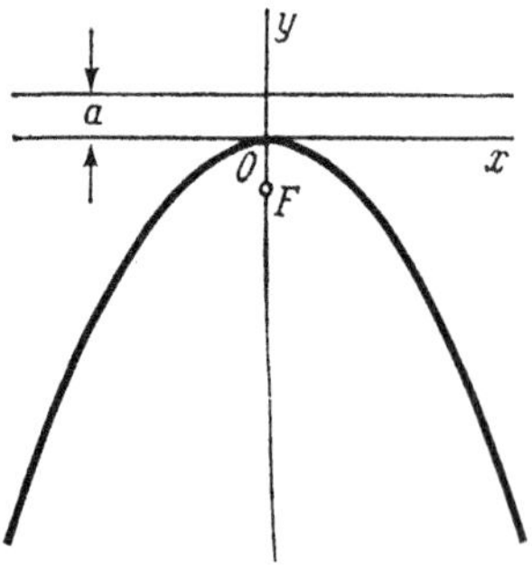

FIG. 22

The equation

$$x^2 = -4ay \tag{4}$$

is the canonical equation of a parabola which has its vertex at the origin, its axis on the y-axis, and which lies entirely in the lower half-plane, as shown in Fig. 22.

583. Set up an equation for a parabola with its vertex at the origin, given that:

1) the parabola lies in the right half-plane, is symmetrical about the x-axis, and its parameter $2a = 3$,

2) the parabola lies in the left half-plane, is symmetrical about the x-axis, and its parameter $2a = 0{\cdot}5$,

3) the parabola lies in the upper half-plane, is symmetrical about the y-axis, and its parameter $2a = 1/4$,

4) the parabola lies in the lower half-plane, is symmetrical about the y-axis, and its parameter $2a=3$.

584. Find the parameter of the parabola given by each of the following equations, and state how the curves are situated relative to the axes:

1) $y^2 = 6x$ 2) $x^2 = 5y$ 3) $y^2 = -4x$ 4) $x^2 = -y$.

585. Set up an equation for a parabola which has its vertex at the origin and which is symmetric about:

1) the x-axis and which goes through the point $A(9, 6)$,
2) the x-axis and which goes through the point $B(-1, 3)$,
3) the y-axis and which goes through the point $C(1, 1)$,
4) the y-axis and which goes through the point $D(4, -8)$.

586. A steel wire is stretched between two points 20 metres apart on the same horizontal level; it may be assumed that the wire hangs in a curve which is approximately an arc of a parabola. At a horizontal distance of 2 metres from one point of suspension the wire sags 14·4 cm below the horizontal level through its ends. Estimate the sag of the wire below this level at its mid-point.

587. Set up an equation for a parabola which goes through the origin, has its focus at $F(0, -3)$, and whose axis lies on the y-axis.

588. What sort of curves are defined by the following equations? Draw their graphs.

1) $y = +2\sqrt{x}$ 2) $y = +\sqrt{-x}$ 3) $y = -3\sqrt{(-2x)}$
4) $y = -2\sqrt{x}$ 5) $x = +\sqrt{(5y)}$ 6) $x = -5\sqrt{-y}$
7) $x = -\sqrt{(3y)}$ 8) $x = +4\sqrt{-y}$.

589. Find the focus F and an equation for the directrix of the parabola given by $y^2 = 24x$.

590. Calculate the focal radius of a point M on the parabola given by $y^2 = 20x$ if the x-coordinate of M is equal to 7.

591. Calculate the focal radius of a point M on the parabola given by $y^2 = 12x$ if the y-coordinate of M is equal to 6.

592. Find the points on the parabola given by $y^2 = 16x$ whose focal radii are equal to 13.

593. Find an equation for a parabola having its focus at $F(-7, 0)$ and its directrix defined by $x-7=0$.

594. Set up an equation for a parabola given that its vertex is at (α, β), its parameter is equal to $2a$, its axis is parallel to Ox, and that it extends indefinitely in the direction of 1) the positive x-axis, 2) the negative x-axis.

595. Set up an equation for a parabola given that its vertex is at (α, β), its parameter is equal to $2a$, its axis is parallel to Oy, and that it extends indefinitely in the direction of 1) the positive y-axis, 2) the negative y-axis.

596. Show that each of the following equations represents a parabola, find the coordinates of the vertex A, the value of its parameter $2a$, and the equation of its directrix:

1) $y^2 = 4x-8$, 2) $y^2 = 4-6x$,
3) $x^2 = 6y+2$, 4) $x^2 = 2-y$.

597. Show that each of the following equations represents a parabola, and find the coordinates of its vertex A, and the value of its parameter $2a$:

1) $y = x^2/4+x+2$, 2) $y = 4x^2-8x+7$,
3) $y = -x^2/6+2x-7$.

598. Show that each of the following equations represents a parabola, and find the coordinates of its vertex A, and the value of its parameter $2a$:

1) $x = 2y^2 - 12y + 14$, 2) $x = -y^2/4 + y$,
3) $x = -y^2 + 2y - 1$.

599. What sort of curves are defined by the following equations? Draw their graphs.

1) $y = 3 - 4\sqrt{(x-1)}$ 2) $x = -4 + 3\sqrt{(y+5)}$
3) $x = 2 - \sqrt{(6-2y)}$ 4) $y = -5 + \sqrt{(-3x-21)}$.

600. Set up an equation for a parabola, given its focus $F(7, 2)$ and the equation of its directrix $x - 5 = 0$.

601. Set up an equation for a parabola, given its focus $F(4, 3)$ and the equation of its directrix $y + 1 = 0$.

602. Set up an equation for a parabola, given its focus $F(2, -1)$ and the equation of its directrix $x - y - 1 = 0$.

603. Find the focus F of a parabola whose vertex is at $A(6, -3)$ and whose directrix is given by $3x - 5y + 1 = 0$.

604. Find an equation for a parabola whose vertex is at $A(-2, -1)$ and whose directrix is given by $x + 2y - 1 = 0$.

605. Find the points of intersection of the straight line and parabola given by $x + y - 3 = 0$ and $x^2 = 4y$.

606. Find the points of intersection of the straight line and parabola given by $3x + 4y - 12 = 0$ and $y^2 = -9x$.

607. Find the points of intersection of the straight line and parabola given by $3x - 2y + 6 = 0$ and $y^2 = 6x$.

608. How are the straight line and parabola given by each pair of the following equations situated? Does the straight line intersect, touch, or pass outside the parabola?

1) $x-y+2=0,\ y^2=8x$ 2) $8x+3y-15=0,\ x^2=-3y$
3) $5x-y-15=0,\ y^2=-5x.$

609. For what values of the gradient m will the straight line given by $y=mx+2$, 1) intersect the parabola given by $y^2=4x$, 2) touch it, 3) pass outside it?

610. Derive the condition for the straight line given by $y=mx+c$ to touch the parabola given by $y^2=4ax$.

611. Show that one and only one tangent can be drawn with a given gradient $m\neq 0$ to touch the parabola given by $y^2=4ax$.

612. Set up an equation for the tangent at the point $M_1(x_1, y_1)$ on the parabola given by $y^2=4ax$.

613. Find an equation for the straight line which touches the parabola defined by $y^2=8x$ and which is parallel to the straight line given by $2x+2y-3=0$.

614. Find an equation for the straight line which touches the parabola defined by $x^2=16y$ and which is perpendicular to the straight line given by $2x+4y+7=0$.

615. Find an equation for the straight line which touches the parabola defined by $y^2=12x$ and which is parallel to the straight line defined by $3x-2y+30=0$ and calculate the distance d between the two lines.

616. On the parabola given by $y^2=64x$ find the point M_1 which is closest to the straight line given by $4x+3y-14=0$, and calculate the distance from M_1 to this straight line.

617. Find equations for the tangents drawn from $A(2, 9)$ to the parabola given by $y^2 = 36x$.

618. A tangent at the point M_1 to the parabola given by $y^2 = 4ax$ intersects the x-axis in P, and Q is the projection of the point P on to the x-axis. Show that the vertex of the parabola bisects the interval PQ.

619. Set up an equation for the chord joining the points of contact of the tangents drawn from the point $A(5, 9)$ to the parabola given by $y^2 = 5x$.

620. Calculate the distance d from the point $P(-3, 12)$ to the chord joining the points of contact of the tangents drawn from P to the parabola given by $y^2 = 10x$.

621. Find the points of intersection of the ellipse and parabola given by $x^2/100 + y^2/225 = 1$ and $y^2 = 24x$.

622. Find the points of intersection of the hyperbola and parabola given by $x^2/20 - y^2/5 = -1$ and $y^2 = 3x$.

623. Find the points of intersection of the parabolas given by $y = x^2 - 2x + 1$ and $x = y^2 - 6y + 7$.

624. Prove that the tangent at a point M on a parabola makes equal angles with the focal radius FM and with the ray drawn through M parallel to the axis of the parabola in that direction in which the curve opens out indefinitely.

625. A ray of light is projected from the focus of the parabola given by $y^2 = 12x$ so that it makes an acute angle α with Ox. It is known that $\tan \alpha = 3/4$. The ray of light is reflected when it meets the parabola. Set up an equation for the straight line on which the reflected ray lies.

626. Show that two parabolas which have a common axis and a common focus situated between their vertices intersect at right angles.

627. Prove that if two parabolas which have perpendicular axes intersect in four points, then these points lie on a circle.

§ 21. The Polar Equation of the Ellipse, Hyperbola and Parabola

The polar equation has the same general form for an ellipse, for one branch of a hyperbola, and for a parabola. If the pole is taken at a focus, and the polar axis is taken along the axis of the curve in the direction away from the directrix nearer to the chosen focus, then the equation has the form

$$\varrho = \frac{p}{1-\varepsilon\cos\theta} \tag{1}$$

where ϱ, θ are the polar coordinates of an arbitrary point of the curve,

p is the focal parameter, i.e., half the length of the chord drawn through the focus perpendicular to the axis,

and ε is the eccentricity ($\varepsilon = 1$ for a parabola).

628. An ellipse is defined by $x^2/25+y^2/16 = 1$. Set up its polar equation in a system having the polar axis directed along the positive x-axis and the pole at 1) the left-hand focus, 2) the right-hand focus.

629. A hyperbola is defined by $x^2/16-y^2/9 = 1$. Set up the polar equation for its right-hand branch in a system having the polar axis directed along the positive x-axis and the pole at 1) the left-hand focus, 2) the right-hand focus.

630. A hyperbola is defined by $x^2/25 - y^2/144 = 1$. Set up the polar equation for its left-hand branch in a system having the polar axis directed along the positive x-axis and the pole at 1) the left-hand focus, 2) the right-hand focus.

631. A parabola is defined by $y^2 = 6x$. Set up its polar equation in a system having the polar axis directed along the positive x-axis and the pole at the focus.

632. What sort of curves are defined by the following equations?

1) $\varrho = \dfrac{5}{1-0{\cdot}5\cos\theta}$ 2) $\varrho = \dfrac{6}{1-\cos\theta}$ 3) $\varrho = \dfrac{10}{1-1{\cdot}5\cos\theta}$

4) $\varrho = \dfrac{12}{2-\cos\theta}$ 5) $\varrho = \dfrac{5}{3-4\cos\theta}$ 6) $\varrho = \dfrac{1}{3-3\cos\theta}$

633. Show that the equation $\varrho = \dfrac{144}{13-5\cos\theta}$ defines an ellipse and find its semi-axes.

634. Show that the equation $\varrho = \dfrac{18}{4-5\cos\theta}$ defines the right-hand branch of a hyperbola and find its semi-axes.

635. Show that the equation $\varrho = \dfrac{21}{5-2\cos\theta}$ defines an ellipse and set up the polar equation for its directrices.

636. Show that the equation $\varrho = \dfrac{16}{3-5\cos\theta}$ defines the right-hand branch of a hyperbola and find polar equations for its directrices and asymptotes.

637. Find the points whose polar radius is equal to 6 on the ellipse defined by $\varrho = \dfrac{12}{3-(\sqrt{2})\cos\theta}$.

638. Find the points whose polar radius is equal to 3 on the hyperbola defined by $\varrho = \dfrac{15}{3-4\cos\theta}$.

639. On the parabola defined by $\varrho = \dfrac{p}{1-\cos\theta}$ find the points

1) with the smallest polar radius
2) with polar radius equal to the parameter of the parabola.

640. Set up the polar equation for the ellipse given by $\dfrac{x^2}{a^2}+\dfrac{y^2}{b^2}=1$ if the pole is taken at the centre of the ellipse and the polar axis is in the positive direction of Ox.

641. Set up the polar equation for the hyperbola given by $x^2/a^2 - y^2/b^2 = 1$ if the pole is taken at the centre and the polar axis is in the positive direction of Ox.

642. Set up the polar equation for the parabola given by $y^2 = 4ax$ if the pole is taken at the vertex of the parabola and the polar axis is in the positive direction of Ox.

§ 22. Diameters of Curves of the Second Order

It is proved in textbooks of analytical geometry that: The mid-points of any family of parallel chords of a second-order curve lie on a straight line. This straight line is called a diameter of the curve.

A diameter which bisects a particular chord is said to be conjugate to it (the diameter will obviously bisect, and be conjugate to, all chords parallel to the chosen one).

All diameters of an ellipse or of a hyperbola go through the centre of the curve.

If an ellipse is given by the equation $\frac{x^2}{a^2}+\frac{y^2}{b^2}=1$ (1)

then its diameter which is conjugate to chords having the gradient m is defined by

$$y=-\frac{b^2}{a^2m}x. \tag{2}$$

If a hyperbola is given by the equation $\frac{x^2}{a^2}-\frac{y^2}{b^2}=1$ (3)

then its diameter which is conjugate to chords having the gradient m is defined by

$$y=\frac{b^2}{a^2m}x. \tag{4}$$

All diameters of a parabola are parallel to its axis. If the parabola is given by

$$y^2=4ax,$$

then its diameter which is conjugate to chords having the gradient m is given by

$$y=\frac{2a}{m}.$$

If two diameters of an ellipse or hyperbola are such that the first diameter bisects all chords parallel to the second diameter, then the second diameter bisects all chords parallel to the first. Two such diameters are said to be (mutually) conjugate.

If m, m' are the gradients of two conjugate diameters of the ellipse defined by equation (1), then

$$mm'=-\frac{b^2}{a^2}. \tag{5}$$

If m, m' are the gradients of two conjugate diameters of the hyperbola defined by equation (3), then

$$mm'=+\frac{b^2}{a^2}. \tag{6}$$

The diameters of a curve of the second order which are both mutually conjugate and mutually perpendicular are called its principal diameters.

643. Find an equation for the diameter of the ellipse given by $x^2/25+y^2/16=1$ which bisects the chord cut off on the straight line given by $2x-y-3=0$.

644. Set up an equation for the chord of the ellipse given by $x^2/16+y^2/9=1$ which is bisected by the point $A(1, -2)$.

645. Set up equations for the two conjugate diameters of the ellipse given by $x^2+4y^2=1$ of which one is inclined at 45° to Ox.

646. An ellipse is defined by $4x^2+9y^2=1$. Set up equations for its two conjugate diameters, one of which is parallel to the straight line given by $x+2y-5=0$.

647. An ellipse is defined by $x^2+3y^2=1$. Set up equations for its two conjugate diameters, one of which is perpendicular to the straight line given by $3x+2y-7=0$.

648. The figure of an ellipse has been drawn. Using ruler and compasses only, construct its centre.

649. Prove that the axes of an ellipse are its only pair of principal diameters.

650. From the properties of conjugate diameters, prove that any diameter of a circle is a principal diameter.

651. a) An isosceles triangle is inscribed in an ellipse so that one of its vertices coincides with a vertex of the ellipse. Prove that the base of the triangle is parallel to one of the axes of the ellipse.

b) Prove that the sides of a rectangle inscribed in an ellipse are parallel to the axes of the ellipse.

c) The figure of an ellipse has been drawn. Using ruler and compasses only, construct its principal diameters.

652. Prove that the chords joining an arbitrary point of an ellipse to the ends of one of its diameters are parallel to a pair of its conjugate diameters.

653. a) Prove that the sum of the squares of any two conjugate semi-diameters of an ellipse is always equal to the sum of the squares of its semi-axes.

b) Prove that the area of a parallelogram constructed on any two conjugate semi-diameters of an ellipse is always equal to the area of the rectangle constructed on its semi-axes.

654. Set up an equation for the diameter of the hyperbola given by $x^2/5-y^2/4=1$ which bisects the chord cut off on the straight line given by $2x-y+3=0$.

655. A hyperbola is defined by $x^2/3-y^2/7=1$. Set up an equation for its chord which is bisected by the point $A(3, -1)$.

656. A hyperbola is defined by $x^2-4y^2=4$. Set up equations for its two conjugate diameters one of which goes through the point $A(8, 1)$.

657. A hyperbola is defined by $x^2/4-y^2/6=1$. Set up equations for its conjugate diameters which form an angle of 45°.

658. The figure of a hyperbola has been drawn. Using ruler and compasses only, construct its centre.

659. Prove that the axes of a hyperbola are its only pair of principal diameters.

660. The figure of a hyperbola has been drawn. Using ruler and compasses only, construct its principal diameters.

661. A parabola is defined by $y^2 = 12x$. Set up an equation for its diameter which bisects the chord cut off on the straight line given by $3x+y-5=0$.

662. A parabola is defined by $y^2 = 20x$. Set up an equation for its chord which is bisected by the point $A(2, 5)$.

663. Prove that the axis of a parabola is its unique principal diameter.

664. The figure of a parabola has been drawn. Using ruler and compasses only, construct its principal diameter.

CHAPTER 5

SIMPLIFICATION OF THE GENERAL EQUATION OF A CURVE OF THE SECOND ORDER. THE EQUATIONS OF CERTAIN CURVES COMMONLY ENCOUNTERED IN MATHEMATICS AND ALLIED SUBJECTS

§ 23. The Centre of a Curve of Second Order

A curve which in some Cartesian coordinate system is defined by an equation of the second degree is called a second-order curve. The general equation of the second degree (in two variables) is usually written in the form

$$ax^2+2hxy+by^2+2gx+2fy+c = 0. \tag{1}$$

If there is a point in the plane such that all points of a given curve can be arranged in pairs symmetrical about it, then this point is called the centre of the given curve. A curve of the second order which has a unique centre is called a central curve.

The point $S(x_0, y_0)$ is the centre of the curve defined by equation (1) if and only if its coordinates satisfy the equations

$$\begin{aligned} ax_0+hy_0+g &= 0 \\ hx_0+by_0+f &= 0. \end{aligned} \tag{2}$$

Let δ denote the determinant of the equations (2):

$$\delta = \begin{vmatrix} a & h \\ h & b \end{vmatrix}.$$

The quantity δ is composed of the coefficients of the second-degree terms of equation (1), and it is called the discriminant of the second-degree terms.

If $\delta \neq 0$, the equations (2) are compatible and determinate, i.e., they have a unique solution. In this case the coordinates of the centre may be determined from the formulae

$$x_0 = \frac{\begin{vmatrix} h & g \\ b & f \end{vmatrix}}{\begin{vmatrix} a & h \\ h & b \end{vmatrix}}, \quad y_0 = \frac{\begin{vmatrix} g & a \\ f & h \end{vmatrix}}{\begin{vmatrix} a & h \\ h & b \end{vmatrix}}.$$

The inequality $\delta \neq 0$ is the criterion for a central (second-order) curve.

If $S(x_0, y_0)$ is the centre of a second-order curve, then, on transforming the coordinates according to the relations

$$x = \tilde{x} + x_0, \quad y = \tilde{y} + y_0$$

(which correspond to a transfer of the origin to the centre of the curve), the equation of the curve takes the form

$$a\tilde{x}^2 + 2h\tilde{x}\tilde{y} + b\tilde{y}^2 + \tilde{c} = 0,$$

where a, b, h are the same as in equation (1), and $\tilde{c}$ is given by

$$\tilde{c} = gx_0 + fy_0 + c.$$

If $\delta \neq 0$, then $\tilde{c}$ can also be written as $\tilde{c} = \dfrac{\Delta}{\delta}$, where

$$\Delta = \begin{vmatrix} a & h & g \\ h & b & f \\ g & f & c \end{vmatrix}.$$

Δ is called the discriminant of the form on the left-hand side of equation (1).

665. Which of the following equations define a central curve? Which of the curves defined by them have no centre, and which have infinitely many centres?

1) $3x^2-4xy-2y^2+3x-12y-7 = 0$
2) $4x^2+5xy+3y^2-x+9y-12 = 0$
3) $4x^2-4xy+y^2-6x+8y+13 = 0$
4) $4x^2-4xy+y^2-12x+6y-11 = 0$
5) $x^2-2xy+4y^2+5x-7y+12 = 0$
6) $x^2-2xy+y^2-6x+6y-3 = 0$
7) $4x^2-20xy+25y^2-14x+2y-15 = 0$
8) $4x^2-6xy-9y^2+3x-7y+12 = 0$.

666. Show that each of the following equations represents a central curve and find the coordinates of the centre:

1) $3x^2+5xy+y^2-8x-11y-7 = 0$
2) $5x^2+4xy+2y^2+20x+20y-18 = 0$
3) $9x^2-4xy-7y^2-12 = 0$
4) $2x^2-6xy+5y^2+22x-36y+11 = 0$.

667. Show that each of the following equations defines a curve which has infinitely many centres, and set up an equation for the locus of the centres:

1) $x^2-6xy+9y^2-12x+36y+20 = 0$
2) $4x^2+4xy+y^2-8x-4y-21 = 0$
3) $25x^2-10xy+y^2+40x-8y+7 = 0$

668. Show that the following equations define central curves and simplify them by transferring the origin to the centre in each case:

1) $3x^2-6xy+2y^2-4x+2y+1 = 0$
2) $6x^2+4xy+y^2+4x-2y+2 = 0$
3) $4x^2+6xy+y^2-10x-10 = 0$
4) $4x^2+2xy+6y^2+6x-10y+9 = 0$.

669. For what values of m and n does the equation

$$mx^2+12xy+9y^2+4x+ny-13 = 0$$

define: a) a central curve, b) a curve without a centre, c) a curve having infinitely many centres?

670. A curve is given by the equation

$$4x^2-4xy+y^2+6x+1 = 0.$$

For what values of the gradient m will the straight line given by $y = mx$

a) intersect this curve in a single point

b) touch the curve

c) intersect the curve in two points

d) have no point in common with the curve?

671. Set up an equation for the second-order curve which has its centre at the origin, goes through the point $M(6, -2)$, and touches the straight line given by $x-2 = 0$ at the point $N(2, 0)$.

672. Set up an equation for the second-order curve which has its centre at the point $P(1, -2)$, goes through the point $Q(0, -3)$, and touches the x-axis at the origin.

§ 24. Reduction of the Equation of a Central Curve of the Second Order to its Simplest Form

A central curve of the second order is given by the equation

$$ax^2+2hxy+by^2+2gx+2fy+c = 0, \quad \delta = ab-h^2 \neq 0. \quad (1)$$

By transferring the origin to the centre $S(x_0, y_0)$ of this curve, using the transformation formulae

$$x = \tilde{x}+x_0, \quad y = \tilde{y}+y_0,$$

we find that equation (1) becomes

$$a\tilde{x}^2+2h\tilde{x}\tilde{y}+b\tilde{y}^2+\tilde{c}=0$$

where $\tilde{c}=gx_0+fy_0+c=\Delta/\delta$.

The further simplification of equation (2) is accomplished by using the transformation formulae

$$\begin{aligned}\tilde{x}&=\tilde{x}'\cos\alpha-\tilde{y}'\sin\alpha\\ \tilde{y}&=\tilde{x}'\sin\alpha+\tilde{y}'\cos\alpha\end{aligned}\qquad(3)$$

corresponding to a rotation of the axes through an angle α.

If the angle α is chosen so that

$$h\tan^2\alpha-(b-a)\tan\alpha-h=0,\qquad(4)$$

then in the new coordinate system the equation of the curve takes the form

$$\tilde{a}'\tilde{x}'^2+\tilde{b}'\tilde{y}'^2+c'=0\qquad(5)$$

where $\tilde{a}'\neq 0$, $\tilde{b}'\neq 0$.

Note. $\tan\alpha$ can be found from equation (4), and then the values of $\sin\alpha$ and $\cos\alpha$ which occur in equation (3) can be calculated from the trigonometrical relations

$$\sin\alpha=\frac{\tan\alpha}{\pm\sqrt{(1+\tan^2\alpha)}},\quad \cos\alpha=\frac{1}{\pm\sqrt{(1+\tan^2\alpha)}}.$$

Between the coefficients of equations (1) and (5) subsist the important relations:

$$\begin{aligned}\tilde{a}'\tilde{b}'&=ab-h^2\\ \tilde{a}'+\tilde{b}'&=a+b.\end{aligned}$$

By means of these relations the values of $\tilde{a}'$ and $\tilde{b}'$ can be calculated without actually carrying out the transformation of coordinates.

An equation of the second degree is said to be

elliptical if $\delta>0$
hyperbolic if $\delta<0$
parabolic if $\delta=0$.

An equation of a central, second-order curve must be either elliptical or hyperbolic.

Any elliptical equation defines either a finite ellipse, or a degenerate ellipse (i.e., a single point), or an imaginary ellipse (i.e., the equation has no geometrical representation).

Any hyperbolic equation defines either an ordinary hyperbola, or a degenerate hyperbola (i.e., a pair of intersecting straight lines).

673. Determine the type of each of the following equations (i.e., whether elliptical, hyperbolic, or parabolic). Simplify each equation by a transformation to parallel axes. Determine the geometrical shape which each defines, and sketch the curves to show their relation to the old and new axes.

1) $4x^2+9y^2-40x+36y+100 = 0$
2) $9x^2-16y^2-54x-64y-127 = 0$
3) $9x^2+4y^2+18x-8y+49 = 0$
4) $4x^2-y^2+8x-2y+3 = 0$
5) $2x^2+3y^2+8x-6y+11 = 0.$

674. Reduce each of the following equations to its simplest form, determine its type, establish the nature of the curve defined by it, and draw a figure to show the position of the curve relative to the old and new axes.

1) $32x^2+52xy-7y^2+180 = 0$
2) $5x^2-6xy+5y^2-32 = 0$
3) $17x^2-12xy+8y^2 = 0$
4) $5x^2+24xy-5y^2 = 0$
5) $5x^2-6xy+5y^2+8 = 0.$

675. Determine the type of each of the following equations by calculating the discriminant of the second-degree terms.

1) $2x^2+10xy+12y^2-7x+18y-15=0$

2) $3x^2-8xy+7y^2+8x-15y+20=0$

3) $25x^2-20xy+4y^2-12x+20y-17=0$

4) $5x^2+14xy+11y^2+12x-7y+19=0$

5) $x^2-4xy+4y^2+7x-12=0$

6) $3x^2-2xy-3y^2+12y-15=0.$

676. Reduce each of the following equations to its canonical form, determine its type, establish the nature of the curve defined by it, and draw a picture to show the curve and its position relative to all the coordinate systems used in the reduction.

1) $3x^2+10xy+3y^2-2x-14y-13=0$

2) $25x^2-14xy+25y^2+64x-64y-224=0$

3) $4xy+3y^2+16x+12y-36=0$

4) $7x^2+6xy-y^2+28x+12y+28=0$

5) $19x^2+6xy+11y^2+38x+6y+29=0$

6) $5x^2-2xy+5y^2-4x+20y+20=0.$

677. Repeat q. **676** for the following equations.

1) $14x^2+24xy+21y^2-4x+18y-139=0$

2) $11x^2-20xy-4y^2-20x-8y+1=0$

3) $7x^2+60xy+32y^2-14x-60y+7=0$

4) $50x^2-8xy+35y^2+100x-8y+67=0$

5) $41x^2+24xy+34y^2+34x-112y+129=0$

6) $29x^2-24xy+36y^2+82x-96y-91=0$

7) $4x^2+24xy+11y^2+64x+42y+51=0$

8) $41x^2+24xy+9y^2+24x+18y-36=0.$

678. Without carrying out a transformation of coordinates, establish that each of the following equations determines an ellipse, and find the magnitude of the semi-axes.

1) $41x^2+24xy+9y^2+24x+18y-36=0$
2) $8x^2+4xy+5y^2+16x+4y-28=0$
3) $13x^2+18xy+37y^2-26x-18y+3=0$
4) $13x^2+10xy+13y^2+46x+62y+13=0.$

679. Without carrying out a transformation of coordinates, establish that each of the following equations defines a single point (a degenerate ellipse), and find its coordinates.

a) $5x^2-6xy+2y^2-2x+2=0$
b) $x^2+2xy+2y^2+6y+9=0$
c) $5x^2+4xy+y^2-6x-2y+2=0$
d) $x^2-6xy+10y^2+10x-32y+26=0.$

680. Without carrying out a transformation of coordinates, establish that each of the following equations defines a hyperbola, and find the magnitude of the semi-axes.

1) $4x^2+24xy+11y^2+64x+42y+51=0$
2) $12x^2+26xy+12y^2-52x-48y+73=0$
3) $3x^2+4xy-12x+16=0$
4) $x^2-6xy-7y^2+10x-30y+23=0.$

681. Without carrying out a transformation of coordinates, establish that each of the following equations defines a pair of intersecting straight lines (a degenerate hyperbola), and find their equations.

a) $3x^2+4xy+y^2-2x-1=0$
b) $x^2-6xy+8y^2-4y-4=0$
c) $x^2-4xy+3y^2=0$
d) $x^2+4xy+3y^2-6x-12y+9=0.$

682. Without carrying out a transformation of coordinates, establish what sort of curves are defined by the following equations.

1) $8x^2-12xy+17y^2+16x-12y+3=0$
2) $17x^2-18xy-7y^2+34x-18y+7=0$
3) $2x^2+3xy-2y^2+5x+10y=0$
4) $6x^2-6xy+9y^2-4x+18y+14=0$
5) $5x^2-2xy+5y^2-4x+20y+20=0$.

683. Prove that in an elliptical equation neither of the coefficients a and b can vanish and that they must have the same sign.

684. Prove that an elliptical equation of the second degree ($\delta > 0$) defines an ellipse if and only if a and Δ are numbers of opposite sign.

685. Prove that an elliptical equation of the second degree ($\delta > 0$) defines an imaginary ellipse if and only if a and Δ are numbers of the same sign.

686. Prove that an elliptical equation of the second degree ($\delta > 0$) defines a degenerate ellipse (a point) if and only if $\Delta = 0$.

687. Prove that a hyperbolic equation of the second degree ($\delta < 0$) defines a hyperbola if and only if $\Delta \neq 0$.

688. Prove that a hyperbolic equation of the second degree ($\delta < 0$) defines a degenerate hyperbola (a pair of intersecting lines) if and only if $\Delta = 0$.

§ 25. Reduction of a Parabolic Equation to its Simplest Form

If the equation

$$ax^2+2hxy+by^2+2gx+2fy+c=0 \tag{1}$$

is parabolic, i.e., if $\delta = ab-h^2 = 0$, then the curve defined by equation (1) either has no centre or has infinitely many centres.

A parabolic equation can be simplified by first rotating the coordinate axes, i.e., by transforming equation (1) using the formulae

$$\left.\begin{aligned} x &= x' \cos\alpha - y' \sin\alpha \\ y &= x' \sin\alpha + y' \cos\alpha \end{aligned}\right\}, \qquad (2)$$

where the angle α is to be found from the equation

$$h \tan^2\alpha - (b-a)\tan\alpha - h = 0. \qquad (3)$$

In the new coordinate system equation (1) takes the form of either

$$a'x'^2 + 2g'x' + 2f'y' + c' = 0 \text{ with } a' \neq 0 \qquad (4)$$

or

$$b'y'^2 + 2g'x' + 2f'y' + c' = 0 \text{ with } b' \neq 0. \qquad (5)$$

An equation of the form (4) or (5) can be further simplified by transferring the rotated axes to a new origin (keeping their directions unchanged).

689. Show that each of the following equations is parabolic. Reduce each to its simplest form, establish the nature of the curve defined by it, and draw a figure to show the curve, the original axes, and any other axes used in the reduction.

1) $9x^2 - 24xy + 16y^2 - 20x + 110y - 50 = 0$
2) $9x^2 + 12xy + 4y^2 - 24x - 16y + 3 = 0$
3) $16x^2 - 24xy + 9y^2 - 160x + 120y + 425 = 0$.

690. Repeat q. **689** for the following equations.

1) $9x^2 + 24xy + 16y^2 - 18x + 226y + 109 = 0$
2) $x^2 - 2xy + y^2 - 12x + 12y - 14 = 0$
3) $4x^2 + 12xy + 9y^2 - 4x - 6y + 1 = 0$.

691. Prove that in any parabolic equation the coefficients a and b cannot be numbers of different sign and that they cannot vanish simultaneously.

692. Prove that any parabolic equation can be written in the form

$$(\alpha x+\beta y)^2+2gx+2fy+c = 0.$$

Prove also that elliptical and hyperbolic equations cannot be written in this form.

693. Show that the following equations are parabolic, and write each of them in the form indicated in q. **692.**

1) $x^2+4xy+4y^2+4x+y-15 = 0$
2) $9x^2-6xy+y^2-x+2y-14 = 0$
3) $25x^2-20xy+4y^2+3x-y+11 = 0$
4) $16x^2+16xy+4y^2-5x+7y = 0$
5) $9x^2-42xy+49y^2+3x-2y-24 = 0.$

694. Show that if an equation of the second degree is parabolic and is written in the form $(\alpha x+\beta y)^2+2gx+2fy+c = 0$, then the discriminant of its left-hand member is given by $\Delta = -(g\beta-f\alpha)^2$.

695. Prove that the parabolic equation

$$(\alpha x+\beta y)^2+2gx+2fy+c = 0$$

under the transformation

$$\begin{aligned} x &= x' \cos\theta - y' \sin\theta \\ y &= x' \sin\theta + y' \cos\theta \end{aligned} \qquad \text{where} \quad \tan\theta = -\frac{\alpha}{\beta}$$

takes the form $b'y'^2+2g'x'+2f'y'+c' = 0,$
where

$$b' = \alpha^2+\beta^2, \quad g' = \pm\sqrt{\left(\frac{-\Delta}{\alpha^2+\beta^2}\right)},$$

Δ being the discriminant of the left-hand member of the given equation.

696. Show that a parabolic equation defines a parabola if and only if $\Delta \neq 0$. Show also that in this case the parameter of the parabola is given by

$$2a = \sqrt{\left(\frac{-\Delta}{(a+b)^3}\right)}$$

697. Without carrying out a transformation of coordinates, show that each of the following equations defines a parabola, and find the parameter of each parabola.

1) $9x^2+24xy+16y^2-120x+90y=0$

2) $9x^2-24xy+16y^2-54yx-178y+181=0$

3) $x^2-2xy+y^2+6x-14y+29=0$

4) $9x^2-6xy+y^2-50x+50y-275=0.$

698. Prove that an equation of the second degree defines a degenerate curve if and only if $\Delta=0$.

699. Without carrying out a transformation of coordinates, show that each of the following equations defines a pair of parallel lines, and find their equations.

a) $4x^2+4xy+y^2-12x-6y+5=0$

b) $4x^2-12xy+9y^2+20x-30y-11=0$

c) $25x^2-10xy+y^2+10x-2y-15=0.$

700. Without carrying out a transformation of coordinates, show that each of the following equations defines a single straight line (a coincident line-pair), and find the equation of this line.

a) $x^2-6xy+9y^2+4x-12y+4=0$

b) $9x^2+30xy+25y^2+42x+70y+49=0$

c) $16x^2-16xy+4y^2-72x+36y+81=0.$

§ 26. The Equations of Certain Curves Commonly Encountered in Mathematics and Allied Subjects

701. Set up an equation for the locus of a point which moves in a plane so that the product of its distances from two given points $F_1(-c, 0)$ and $F_2(c, 0)$ is always equal to a^2. Such a locus is called a Cassini oval. See Fig. 23.

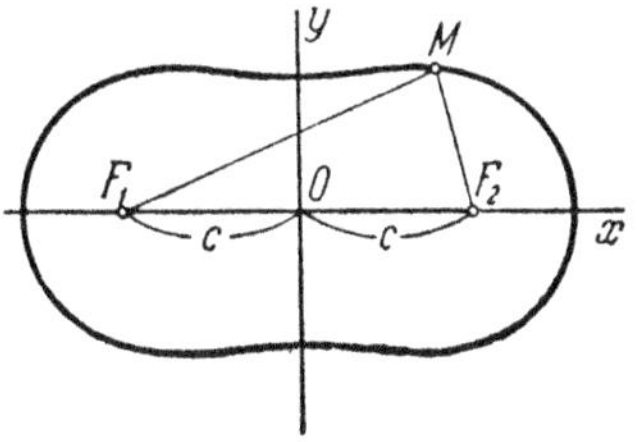

FIG. 23

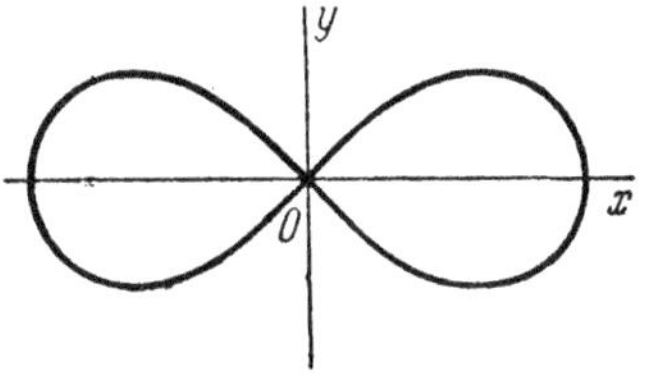

FIG. 24

702. Set up an equation for the locus of a point which moves in a plane so that the product of its distances from two given points $F_1(-a, 0)$ and $F_2(a, 0)$ is always equal to a^2. Such a locus is called a lemniscate (see Fig. 24). (Find an equation for the lemniscate first directly, and then by regarding it as a special case of Cassini's ovals.) Find also the polar equation of a lemniscate, taking the pole at the origin and the polar axis along the positive x-axis.

703. Set up an equation for the locus of the foot of the perpendicular drawn from the origin on to a line which forms with the coordinate axes a triangle of constant area S.

Hint. Find first the polar equation for the locus, taking the pole at the origin and the polar axis along the positive x-axis.

704. Show that the locus discussed in q. **703** is a lemniscate (see q. **702**).

Hint. Rotate the coordinate axes through 45°.

705. A ray a which initially coincides with the polar axis rotates round the pole O with constant angular velocity ω. Set up an equation in the given coordinate system for the path of a point M which, starting from O, moves along the ray a with constant speed v. (This locus is an Archimedean spiral; see Fig. 25.)

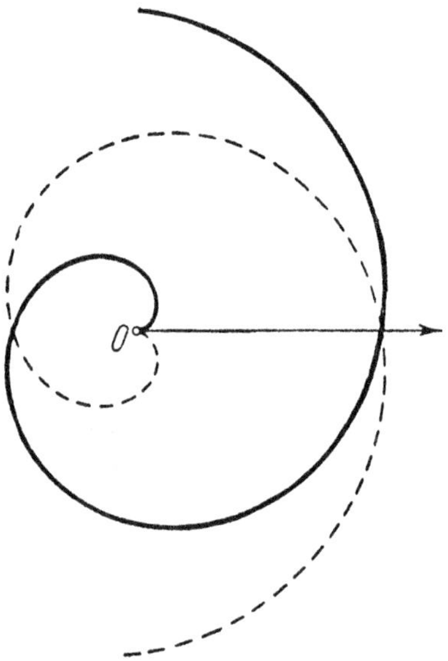

FIG. 25

706. A circle of radius r goes through the origin and touches the straight line given by $x = 2r$. A ray drawn from the origin intersects the circle in the point B and the straight line in the point C. An interval $OM = BC$ is cut off on this ray. As the ray rotates about O the length of the interval OM varies and the point M describes a curve known as a cissoid (see Fig. 26). Find an equation for the cissoid.

707. A circle of diameter a goes through the origin O and touches the straight line given by $x = a$. A ray drawn from the origin intersects the circle in the point A and the straight line in the point B. Straight lines are drawn through A and B, AM parallel to Oy and BM parallel to Ox, to intersect in a point M (see Fig. 27). As the ray rotates about O, the point M describes a curve known as the Witch of Agnesi or the Versiera. Find an equation for this curve.

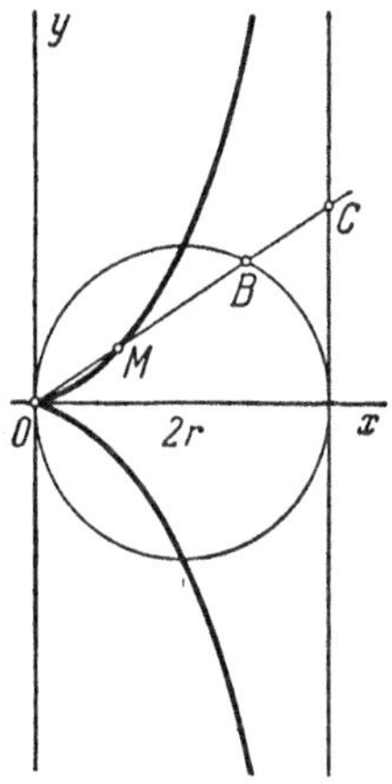

FIG. 26

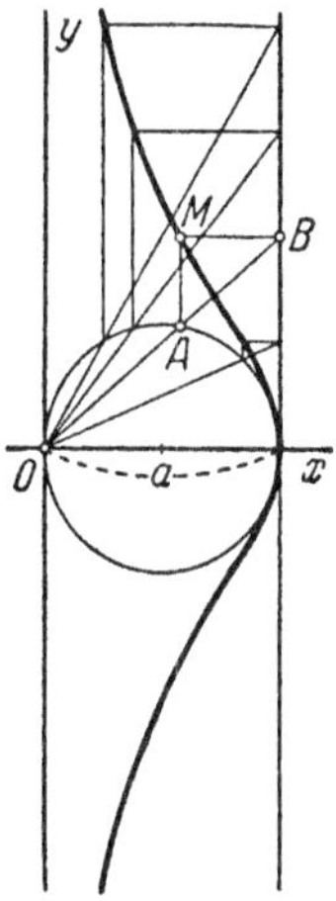

FIG. 27

708. From the point $A(-a, 0)$, where $a > 0$, a ray is drawn to intersect the y-axis in B (see Fig. 28). On either side of the point B intervals BM, BN of equal length b (b = constant) are cut off along the ray. As the ray rotates about A the points M, N describes curves which taken together form the curve known as the conchoid. Find an equation for the conchoid, firstly in polar coordinates, taking the pole at A and the polar axis along the positive x-axis, and then transform it into an equation in Cartesian coordinates for the system shown in the figure.

709. From the point $A(-a, 0)$, where $a > 0$, a ray is drawn to intersect the y-axis in B. On either side of the point B intervals BM, BN of length equal to OB are cut off along the ray. As the ray rotates about A, the points M and N describe a curve known as the strophoid (see Fig. 29). Find an equation for the curve, firstly in polar coordinates, taking the pole at A

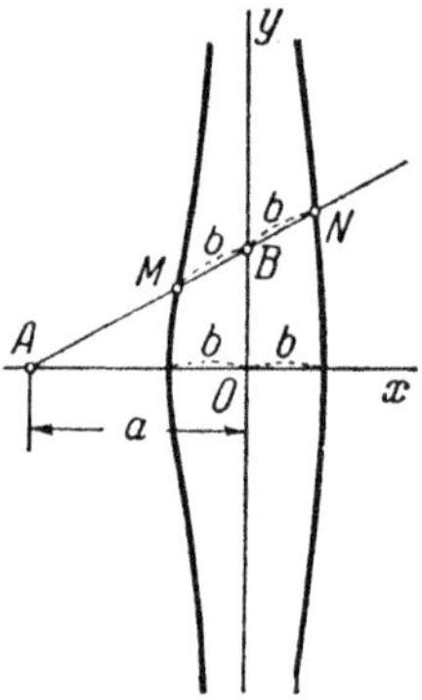

FIG. 28

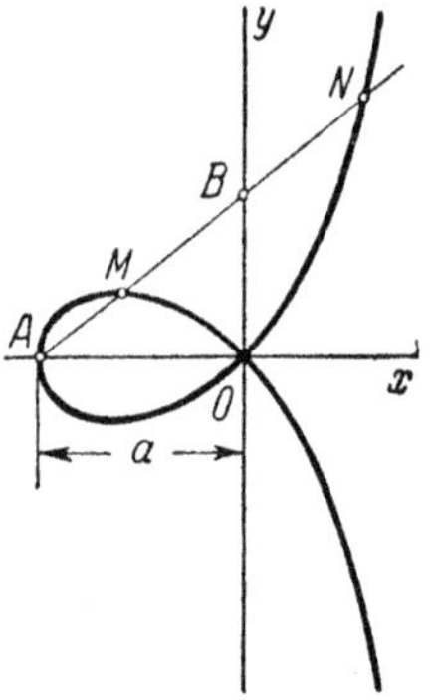

FIG. 29

and the polar axis along the positive x-axis, and then go over to the rectangular Cartesian coordinate system shown in the figure.

710. A ray is drawn from the origin O to intersect in the point B the circle given by $x^2+y^2=2ax$ $(a > 0)$. On either side of the point B intervals BM, BN of constant length b are cut off along the ray. As the ray rotates about O, the points M and N describe a curve known as the limaçon of Pascal. ("Limaçon" because of a fancied resemblance to a snail. The curve was discovered by Etienne Pascal, father of Blaise.) Find an equation for the curve, firstly in polar coordinates, taking the pole at the origin, and the polar axis along the positive

x-axis; and then go over to the given rectangular Cartesian coordinate system. (See Fig. 30.)

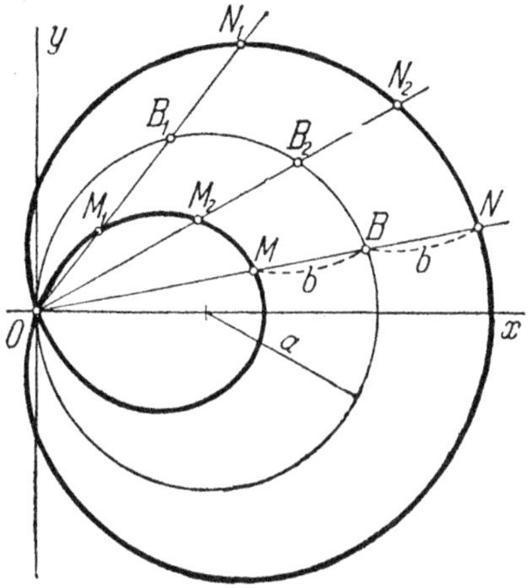

FIG. 30

711. An interval of length $2a$ moves so that its ends are always on the coordinate axes. M is the foot of the perpendicular from the origin on to this interval. The locus traced out

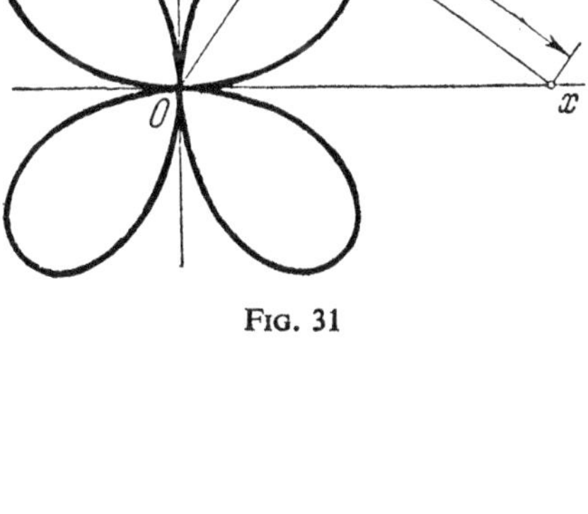

FIG. 31

by M is known as the quadrifolium (or four-petalled rose). Find an equation for the curve, firstly in polar coordinates, taking the pole at the origin and the polar axis along the positive x-axis, and then go over to the given rectangular Cartesian coordinate system. (See Fig. 31.)

712. An interval of length a moves so that its ends are always on the coordinate axes. P is the point of intersection of the straight lines drawn through the ends of the interval parallel to the axes (see Fig. 32). M is the foot of the perpendicular from P on to the interval. The locus of the point M as the interval moves is called an astroid. Find an equation for the curve.

Hint. First set up parametric equations for the locus, taking as parameter the acute angle t between the interval and the x-axis. Then eliminate t.

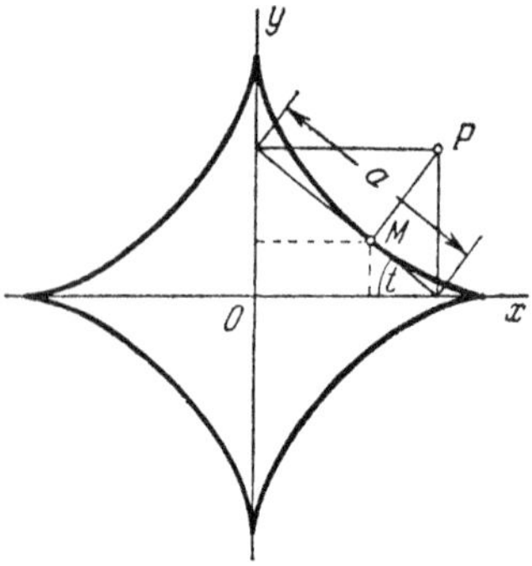

FIG. 32

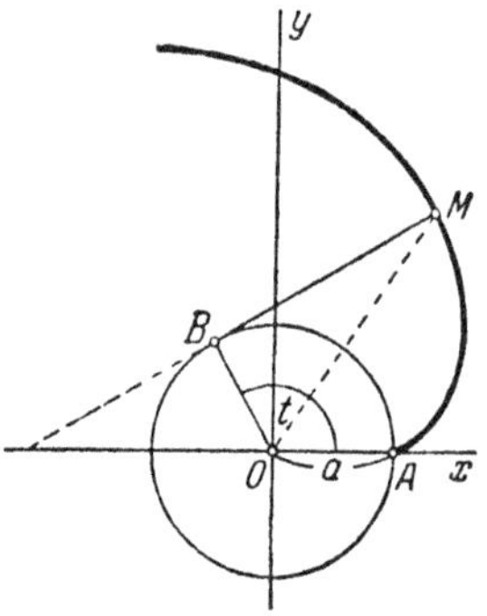

FIG. 33

713. A ray OB is drawn to intersect at B the circle given by $x^2+y^2 = ax$. C is the projection of B on to the x-axis. From C a perpendicular CM is dropped on to the ray OB. The curve traced out by M as the ray OB rotates is called a simple folium.

Find an equation for the curve, first in polar coordinates, taking the pole at the origin and the polar axis along the positive x-axis, and then go over to the given rectangular Cartesian coordinate system.

714. Imagine a thread wound round the circle given by $x^2+y^2=a^2$. The free end of the thread, M, is initially at the point $A(a, 0)$, where $a > 0$. The thread is unwound keeping the free part always taut, so that BM in Fig. 33 is a tangent to the circle at the point B where the thread is just about to leave the circle. The locus of the point M is called the involute of a circle. Find parametric equations for this curve, taking as parameter t the angle AOB.

715. A circle of radius a rolls without slipping along the axis Ox. The path described by a point M on the circumference

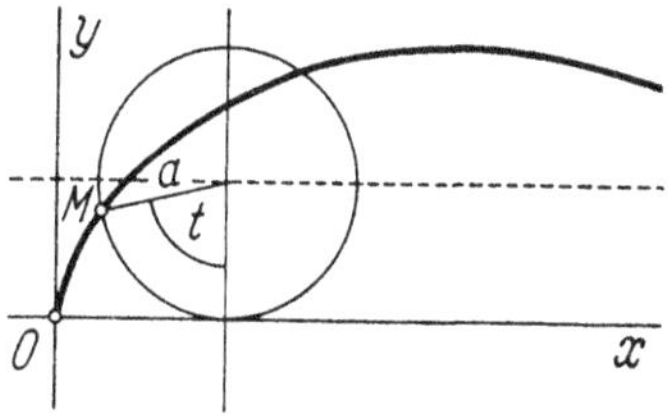

FIG. 34

of the circle is called a cycloid. Derive parametric equations for the curve, taking as parameter the angle t through which the circle has rotated and supposing that initially (when $t=0$) the point M is at the origin. Then eliminate the parameter t (Fig. 34).

716. A circle of radius a rolls without slipping round the outside of the circle given by $x^2+y^2=a^2$. The path described by a point M on the circumference of the rolling circle is called a cardioid (Fig. 35). t is the angle of inclination to the x-axis of the radius of the fixed circle that passes through the point of contact of the moving circle. Derive parametric equations for the cardioid, taking t as parameter and supposing that initially, when $t = 0$, the point M is on the x-axis and to the right of the centre of the fixed circle. Then find the polar equation for the curve, taking the pole at the point A and the polar axis in the direction of the positive x-axis. Show also that the cardioid is a special case of the limaçon of Pascal.

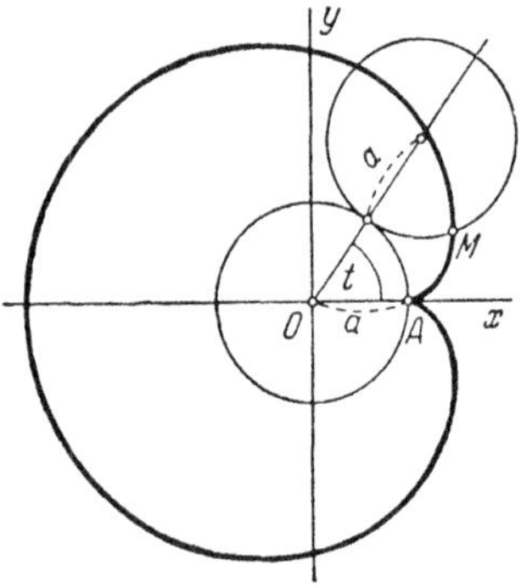

FIG. 35

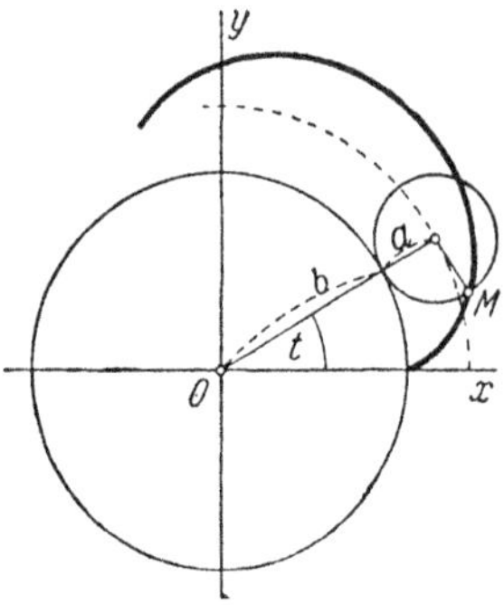

FIG. 36

717. A circle of radius a rolls without slipping round the outside of the fixed circle given by $x^2+y^2=b^2$. The path described by a point M on the circumference of the rolling circle is called an epicycloid. t is the angle which the radius of the fixed circle passing through the point of contact of the moving circle makes with the x-axis (see Fig. 36). Derive parametric equations for the epicycloid, taking t as parameter and supposing that initially, when $t = 0$, the point M is on the x-axis and to the right of the centre of the fixed circle. Show also that the cardioid (see q. **716**) is a special case of an epicycloid.

718. A circle of radius a rolls without slipping round the inside of the circumference of a fixed circle given by $x^2+y^2 = b^2$. The path described by a point M on the circumference of the rolling circle is called a hypocycloid. t is the angle of inclination to the x-axis of the radius of the fixed circle that passes through the point of contact of the moving circle (see

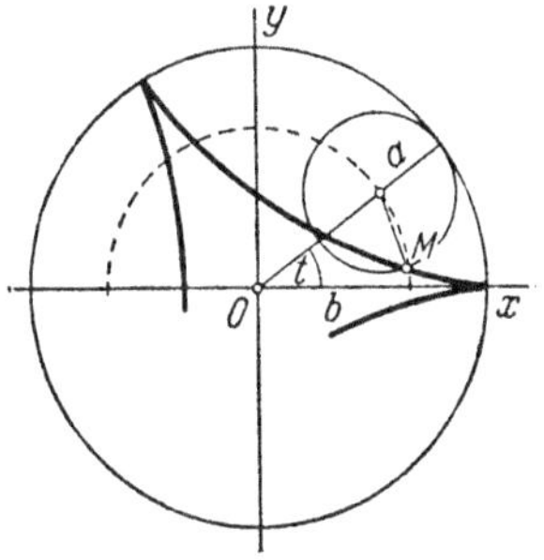

FIG. 37

Fig. 37). Derive parametric equations for the hypocycloid, taking t as parameter, and supposing that initially, when $t=0$, the point M is on the x-axis and to the right of the centre of the fixed circle. Show also that the astroid (see q. **712**) is a special case of a hypocycloid.

(Figures 38–53 appear in the text of Part II, which comprises Questions 719–1261.)

ANSWERS TO QUESTIONS AND NOTES ON SOLUTIONS

PART I

1. See Fig. 54a. **2.** *Note.* The equation $|x| = 2$ is equivalent to the two equations $x = -2$ and $x = 2$ to which correspond the two points $A_1(-2)$ and $A_2(2)$, see Fig. 54b. Similarly the equation $|x-1| = 3$ is equivalent to the two equations $x-1 = -3$ and $x-1 = 3$, giving the two points B_1 and B_2 (Fig. 54b). The other parts are solved similarly.

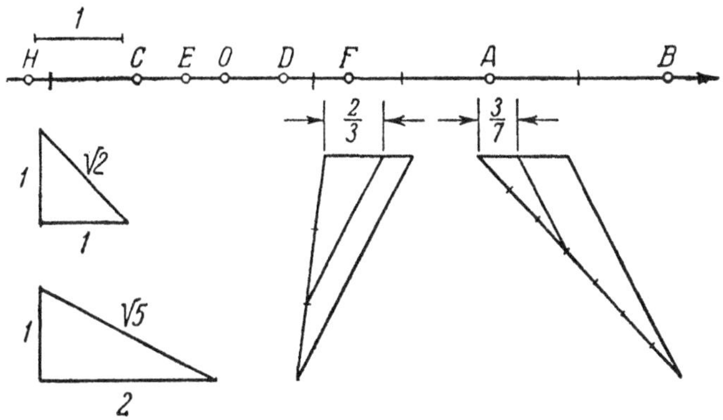

FIG. 54a

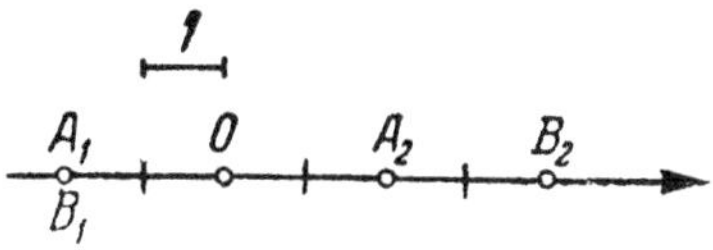

FIG. 54b

3. The points are situated: 1) to the right of the point $M_1(2)$; 2) to the left of the point $M_2(3)$ and include M_2 itself; 3) to the right of the point $M_3(12)$; 4) to the left of the point $M_4(3/2)$ and include M_4 itself; 5) to the right of the point $M_5(5/3)$; 6) within the interval bounded by the points $M_6(1)$ and $M_2(3)$; 7) within the interval bounded by the points $M_7(-2)$ and $M_2(3)$, and include the points M_7 and M_2; 8) within the interval bounded by the points $A(1)$ and $B(2)$; 9) outside the interval bounded by the points $P(-1)$ and $Q(2)$; 10) outside the interval bounded by the points $A(1)$ and $B(2)$; 11) inside the interval bounded by the points $P(-1)$ and $Q(2)$; 12) inside the interval bounded by the points $M(3)$ and $N(5)$, and include the points M and N; 13) outside the interval bounded by the points $M(3)$ and $N(5)$; 14) outside the interval bounded by the points $P_1(-4)$ and $Q_1(3)$; 15) inside the interval bounded by the points $P_1(-4)$ and $Q_1(3)$, and include the points $P_1(-4)$ and $Q_1(3)$. **4.** 1) $AB = 8$, $|AB| = 8$; 2) $AB = -3$, $|AB| = 3$; 3) $AB = 4$, $|AB| = 4$; 4) $AB = 2$, $|AB| = 2$; 5) $AB = -2$, $|AB| = 2$; 6) $AB = 2$, $|AB| = 2$. **5.** 1) -2; 2) 5; 3) 1; 4) -8; 5) -2 and 2; 6) -1 and 5; 7) -6 and 4; 8) -7 and -3. **6.** 1) inside the interval bounded by the points $A(-1)$ and $B(1)$;

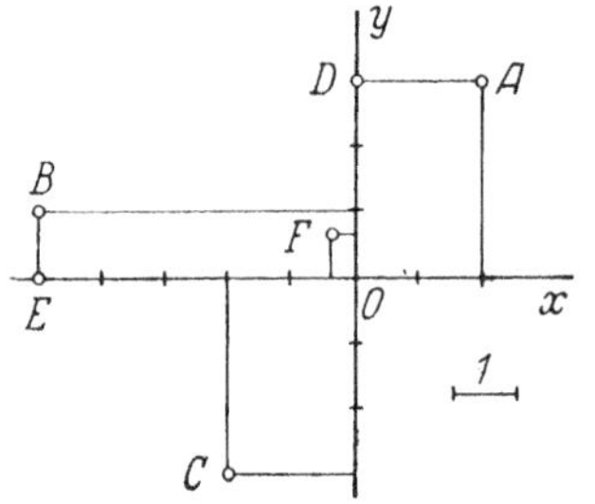

FIG. 55

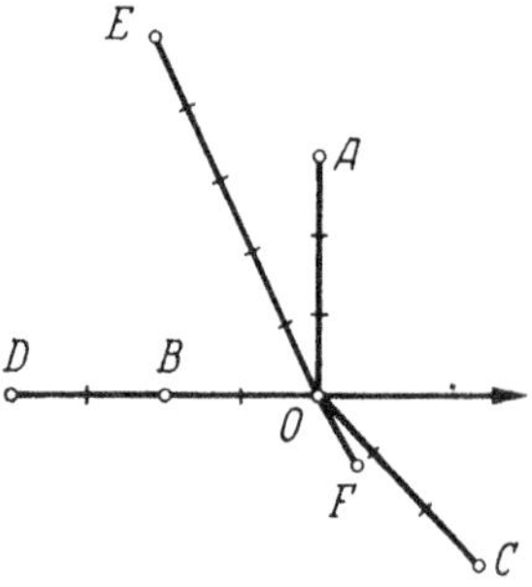

FIG. 56

2) outside the interval bounded by the points $A(-2)$ and $B(2)$; 3) inside the interval bounded by the points $A(-2)$ and $B(2)$, and including the points A and B; 4) outside the interval bounded by the points $A(-3)$ and $B(3)$ and including the points A and B; 5) inside the interval bounded by the points $A(-1)$ and $B(5)$; 6) inside the interval bounded by the points $A(4)$ and $B(6)$, and including the points A and B; 7) outside the interval bounded by the points $A(-1)$ and $B(3)$ and including the points A and B; 8) outside the interval bounded by the points $A(2)$ and $B(4)$, and including the points A and B; 9) inside the interval bounded by the points $A(-4)$ and $B(2)$; 10) outside the interval bounded by the points $A(-3)$ and $B(-1)$; 11) inside the interval bounded by the points $A(-6)$ and $B(-4)$, including the points A and B; 12) outside the interval bounded by the points $A(-3)$ and $B(1)$, and including the points A and B. **7.** 1) 1; 2) $-5/3$; 3) 2; 4) $1/2$; 5) $-10/3$. **8.** $\lambda_1 = AB/BC = 3$; $\lambda_2 = CB/BA = 1/3$; $\lambda_3 = AC/CB = -4$; $\lambda_4 = BC/CA = -1/4$; $\lambda_5 = BA/AC = -3/4$; $\lambda_6 = CA/AB = 4/3$. **9.** $\lambda = (x - x_1)/(x_2 - x)$. **10.** $x = (x_1 + \lambda x_2)/(1 + \lambda)$. **11.** $x = (x_1 + x_2)/2$. **12.** 1) 4; 2) 2; 3) -2; 4) -2; 5) $-1/2$. **13.** 1) $17/3$; 2) $-13/4$; 3) $1/3$; 4) 7; 5) 3; 6) 0. **14.** 1) (-11); 2) $N(13)$. **15.** (5) and (12). **16.** $A(7)$, $B(-41)$. **17.** See Fig. 55. **18.** $A_x(2, 0)$, $B_x(3, 0)$, $C_x(-5, 0)$, $D_x(-3, 0)$, $E_x(-5, 0)$. **19.** $A_y(0, 2)$, $B_y(0, 1)$,

$C_y(0, -2)$, $D_y(0, 1)$, $E_y(0, -2)$. **20.** $A'(2, -3)$, $B'(-3, -2)$, $C'(-1, 1)$, $D'(-3, 5)$, $E'(-4, -6)$, $F'(a, -b)$. **21.** 1) $(1, 2)$, 2) $(-3, -1)$, 3) $(2, -2)$, 4) $(2, 5)$, 5) $(-3, -5)$, 6) $(-a, b)$. **22.** 1) $(-3, -3)$, 2) $(-2, 4)$, 3) $(2, -1)$, 4) $(-5, 3)$, 5) $(5, 4)$, 6) $(-a, -b)$. **23.** 1) $(3, 2)$, 2) $(-2, 5)$, 3) $(4, -3)$. **24.** 1) $(-5, -3)$, 2) $(-3, 4)$, 3) $(2, -7)$. **25.** 1) 1st and 3rd, 2) 2nd and 4th, 3) 1st and 3rd, 4) 2nd and 4th, 5) 1st, 2nd and 4th, 6) 2nd, 3rd and 4th, 7) 1st, 3rd and 4th, 8) 1st, 2nd and 3rd. **26.** See Fig. 56. **27.** $(3, -\pi/4)$, $(2, \pi/2)$, $(3, \pi/3)$, $(1, -2)$, $(5, 1)$. **28.** $(1, -3\pi/4)$, $(5, -\pi/2)$, $(2, 2\pi/3)$, $(4, -\pi/6)$, $(3, \pi-2)$. **29.** $C(3, 5\pi/9)$, $D(5, -11\pi/14)$. **30.** $(1, -2\pi/3)$. **31.** $A(3, -\pi/2)$, $B(2, 3\pi/4)$, $C(1, 0)$, $D(5, \pi/4)$, $E(3, 2-\pi)$, $F(2, \pi-1)$. **32.** $M_1(3, 0)$, $M_2(1, \pi/3)$, $M_3(2, -\pi/3)$, $M_4(5, -\pi/12)$, $M_5(3, \pi)$, $M_6(1, 7\pi/12)$. **33.** $(6, \pi/9)$. **34.** $d = \sqrt{[\varrho_1^2+\varrho_2^2-2\varrho_1\varrho_2 \cos(\theta_2-\theta_1)]}$. **35.** $d = 7$. **36.** $9(17-4\sqrt{3})$ units of area. **37.** $2(13+6\sqrt{2})$ units of area. **38.** $28\sqrt{3}$ units of area. **39.** $S = \frac{1}{2}\varrho_1\varrho_2 \sin(\theta_1-\theta_2)$. **40.** 5 units of area. **41.** $3(4\sqrt{3}-1)$ units of area. **42.** $M_1(0, 6)$, $M_2(5, 0)$, $M_3(\sqrt{2}, \sqrt{2})$, $M_4(5, -5\sqrt{3})$, $M_5(-4, 4\sqrt{3})$, $M_6(6\sqrt{3}, -6)$. **43.** $M_1(5, \pi/2)$, $M_2(3, \pi)$, $M_3(2, \pi/6)$, $M_4(2, -3\pi/4)$, $M_5(2, -\pi/3)$. **44.** 1) 3, 2) -3, 3) 0, 4) 5, 5) -5, 6) 2. **47.** 1) $X = 1$, $Y = 3$, 2) $X = -4$, $Y = -2$, 3) $X = 1$, $Y = -7$, 4) $X = 5$, $Y = 3$. **48.** $(3, -1)$. **49.** $(-3, 2)$. **52.** 1) $X = -6$, $Y = 6\sqrt{3}$, 2) $X = 3\sqrt{3}$, $Y = -3$, 3) $X = \sqrt{2}$, $Y = -\sqrt{2}$. **53.** 1) 5, 2) 13, 3) 10. **54.** 1) $d = 2$, $\theta = \pi/3$, 2) $d = 6$, $\theta = -\pi/4$, 3) $d = 4$, $\theta = 5\pi/6$. **55.** a) $d = \sqrt{2}$, $\theta = -3\pi/4$, b) $d = 5$, $\theta = \tan^{-1} 4/3-\pi$, c) $d = 13$, $\theta = \pi-\tan^{-1} 12/5$, d) $d = \sqrt{234}$, $\theta = -\tan^{-1} 5$. **56.** a) 3, b) -3. **57.** a) $(-9, 3)$, b) $(-9, -7)$. **58.** a) $(-15, -12)$, b) $(1, -12)$. **59.** -2. **60.** $(3\sqrt{3}-4)/2$. **61.** 4. **62.** 1) -5, 2) 5. **63.** 1) 5, 2) 10, 3) 5, 4) $\sqrt{5}$, 5) $2\sqrt{2}$, 6) 13. **64.** 137 units of area. **65.** 34 units of area. **66.** $8\sqrt{3}$ units of area. **67.** 13, 15. **68.** 150 units of area. **69.** $4\sqrt{2}$. **73.** $\angle M_2M_1M_3$ is obtuse. **75.** $\angle BAC = 45°$, $\angle ABC = 45°$, $\angle ACB = 90°$. **76.** 60°. *Note.* Calculate the lengths of the sides of the triangle and then use the cosine formula. **77.** $M_1(6, 0)$ or $M_2(-2, 0)$. **78.** $M_1(0, 28)$ or $M_2(0, -2)$. **79.** $P_1(1, 0)$ or $P_2(6, 0)$. **80.** $C_1(2, 2)$, $R_1 = 2$ or $C_2(10, 10)$, $R_2 = 10$. **81.** $C_1(-3, -5)$ or $C_2(5, -5)$. **82.** $M_2(3, 0)$. **83.** $B(0, 4)$, $D(-1, -3)$. **84.** The conditions of the problem are satisfied by two squares, symmetrically situated relative to the side AB. The other vertices of one of these squares are $C_1(-5, 0)$, $D_1(-2, -4)$, and of the other are $C_2(3, 6)$, $D_2(6, 2)$. **85.** $C(3, -2)$, $R = 10$. **86.** $(1, -2)$. **87.** $Q(4, 6)$. **88.** The mid-points of the sides AB, BC, CA are respectively $(2, -4)$, $(-1, 1)$, $(-2, 2)$. **89.** 1) $M(1, 3)$, 2) $N(4, -3)$. **90.** $(1, -3)$, $(3, 1)$, $(-5, 7)$. **91.** $D(-3, 1)$. **92.** $(5, -3)$, $(1, -5)$. **93.** $D_1(2, 1)$, $D_2(-2, 9)$, $D_3(6, -3)$. *Note:* the 4th vertex of the parallelogram can be opposite any one of the given three points, and so the conditions of the problem are satisfied by three parallelograms. **94.** 13. **95.** $(2, -1)$, $(3, 1)$. **96.** $(5/2, -2)$. **97.** $14\sqrt{2}/3$. **98.** $(-11, -3)$. **99.** 4. **100.** $\lambda_1 = AB/BC = 2$, $\lambda_2 = AC/CB = -3$, $\lambda_3 = BA/AC = -2/3$. **101.** $A(3, -1)$, $B(0, 8)$. **102.** $(3, -1)$. **103.** $(4, -5)$. **104.** $(-9, 0)$. **105.** $(0, -3)$.

106. 1:3 reckoned from B. **107.** (9/2, 1). **108.** $x=(x_1+x_2+x_3)/3$, $y=(y_1+y_2+y_3)/3$. **109.** $M(-1, 0)$, $C(0, 2)$. **111.** (5, 5). **112.** $(5a/12, 5b/12)$. **113.** $(19a/21, 19a/21)$. **114.** $x=(mx_1+nx_2+px_3)/(m+n+p)$, $y=(my_1+ny_2+py_3)/(m+n+p)$. **115.** (4, 2). *Note:* the weight of a uniform wire is proportional to its length. **116.** 1) 14, 2) 12, 3) 26 units of area. **117.** 5. **118.** 20 units of area. **119.** 7·4. **120.** $x=-6/11$, $y=45/11$. **121.** $x=7/17$, $y=10/3$. **122.** (0, −8) or (0, −2) **123.** (5, 0) or (−1/3, 0). **124.** (5, 2) or (2, 2). **125.** $C_1(-7, -3)$, $D_1(-6, -4)$ or $C_2(17, -3)$, $D_2(18, -4)$. **126.** $C_1(-2, 12)$, $D_1(-5, 16)$ or $C_2(-2, 2/3)$, $D_2(-5, 14/3)$. **127.** 1) $x=\tilde{x}+3, y=\tilde{y}+4$, 2) $x=\tilde{x}-2$, $y=\tilde{y}+1$, 3) $x=\tilde{x}-3$, $y=\tilde{y}+5$. **128.** $A(4, -1)$, $B(0, -4)$, $C(2, 0)$. **129.** 1) $A(0, 0)$, $B(-3, 2)$, $C(-4, 4)$, 2) $A(3, -2)$, $B(0, 0)$, $C(-1, 2)$, 3) $A(4, -4)$, $B(1, -2)$, $C(0, 0)$. **130.** 1) (3, 5), 2) (−2, 1), 3) (0, −1), 4) (−5, 0). **131.** 1) $x=(x'-y'\sqrt{3})/2$, $y=(x'\sqrt{3}+y')/2$, 2) $x=(x'+y')/\sqrt{2}$, $y=(-x'+y')/\sqrt{2}$, 3) $x=-y'$, $y=x'$. 4) $x=y'$, $y=-x'$, 5) $x=-x'$, $y=-y'$. **132.** $A(3\sqrt{3}, 1)$, $B(\sqrt{3}/2, 3/2)$, $C(3, -\sqrt{3})$. **133.** 1) $M(\sqrt{2}, 2\sqrt{2})$, $N(-3\sqrt{2}, 2\sqrt{2})$, $P(-\sqrt{2}, -2\sqrt{2})$, 2) $M(1, -3)$, $N(5, 1)$, $P(-1, 3)$, 3) $M(-1, 3)$, $N(-5, -1)$, $P(1, -3)$, 4) $M(-3, -1)$, $N(1, -5)$, $P(3, 1)$. **134.** 1) 60°, 2) −30°. **135.** $\tilde{O}(2, -4)$. **136.** $x=\tilde{x}+1$, $y=\tilde{y}-3$. **137.** $x=3x'/5+4y'/5$, $y=-4x'/5+3y'/5$. **138.** $M_1(1, 5)$, $M_2(2, 0)$, $M_3(16, -5)$. **139.** $A(6, 3)$, $B(0, 0)$, $C(5, -10)$. **140.** 1) $\tilde{O}'(3, -2)$, $\alpha=90°$, 2) $\tilde{O}'(-1, 3)$, $\alpha=180°$, 3) $\tilde{O}'(5, -3)$, $\alpha=-45°$. **141.** $x=-15\tilde{x}'/17-8\tilde{y}'17+9$, $y=8\tilde{x}'/17-15\tilde{y}'/17-3$. **142.** $M_1(1, 9)$, $M_2(4, 2)$, $M_3(1, -3)$, $M_4(0, 2+\sqrt{3})$, $M_5(1+\sqrt{3}, 1)$. **143.** $M_1(0, 5)$, $M_2(3, 0)$, $M_3(-1, 0)$, $M_4(0, -6)$, $M_5(\sqrt{3}, 1)$. **144.** $M_1(2, 0)$, $M_2(1, -\pi/2)$, $M_3(3, \pi/2)$, $M_4(2, -\pi/4)$, $M_5(2, \pi/6)$. **145.** $M_1(\sqrt{2}, \pi/2)$, $M_2(2, -\pi/2)$, $M_3(2, \pi/12)$, $M_4(2, 7\pi/12)$, $M_5(4, -5\pi/12)$. **146.** $f(x, y)=2ax-a^2$. **147.** 1) $f(x, y)=2ax$, 2) $f(x, y)=-2ax-a^2$. **148.** $f(x, y)=4x^2+4y^2+2a^2$. **149.** $f(x, y)=4x^2+4y^2-4ax-4ay+4a^2$. **150.** $f(x, y)=x^2+y^2-25$. **151.** $f(x, y)=2xy-16$. **152.** $f(x, y)$ does not change when the axes are rotated. **153.** (3, 1). **154.** There is no such point. **155.** ±45° or ±135°. **156.** 30°, 120°, −60°, −150°. **157.** M_1, M_4, M_5 lie on the curve. The equation defines the bisector of the angle between the coordinate axes that lies in the 2nd and 4th quadrants (see Fig. 57). **158.** a) (0, −5), (0, 5), b) (−3, −4), (−3, 4), c) (5, 0), d) there are no such points on the given curve, e) (−4, 3), (4, 3), f) (0, −5), g) there are no such points on the given curve. The equation defines a circle, centre $O(0, 0)$, and radius 5 (see Fig. 58). **159.** 1) Bisector of angle between coordinate axes lying in the 1st and 3rd quadrants. 2) Bisector of angle between axes lying in the 2nd and 4th quadrants. 3) The straight line parallel to Oy which cuts off on the positive x-axis an interval equal to 2, measured from the origin (see Fig. 59). 4) The straight line parallel to Oy which cuts off on the negative x-axis an interval equal to 3, measured from the origin (Fig. 59). 5) The straight line parallel to Ox which cuts off on the positive y-axis an interval equal to 5, measured from the origin (Fig. 59). 6) The straight line parallel to Ox which cuts off on the negative y-axis an interval equal to 2, measured from the origin. 7) The y-axis. 8) The x-axis. 9) The curve consists of 2 straight lines: the y-axis itself, and the bisector

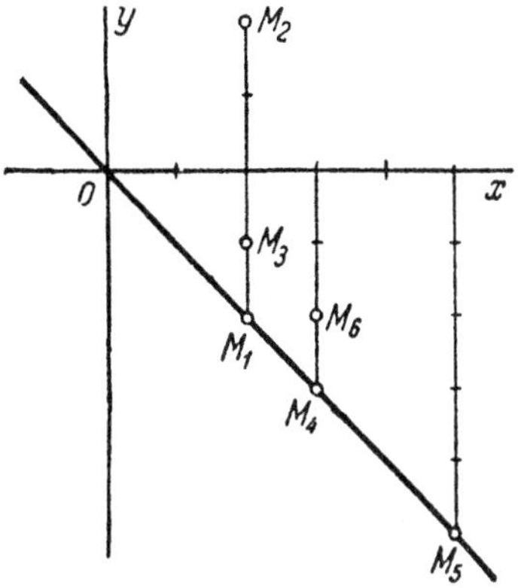

FIG. 57

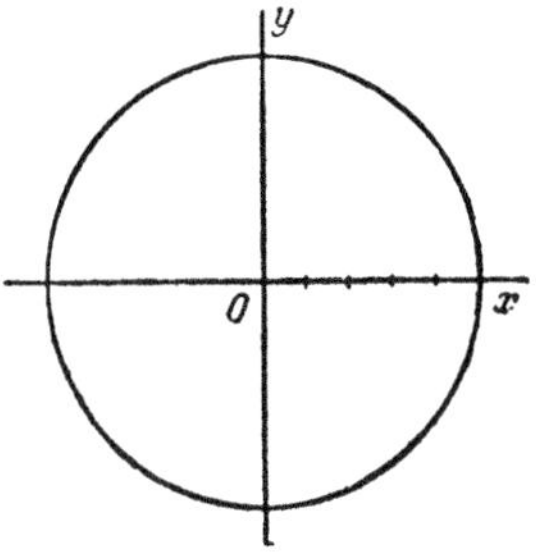

FIG. 58

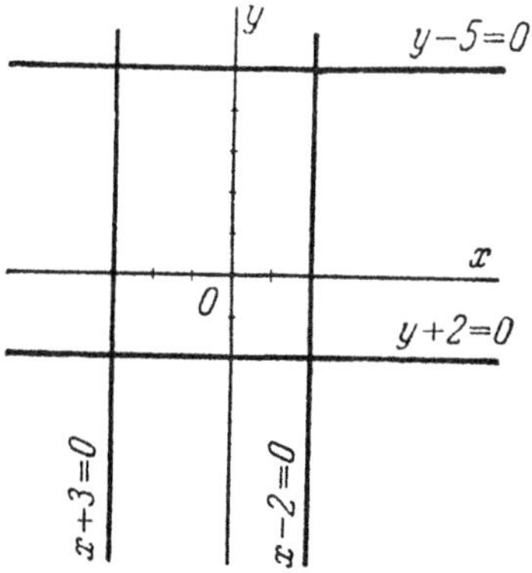

FIG. 59

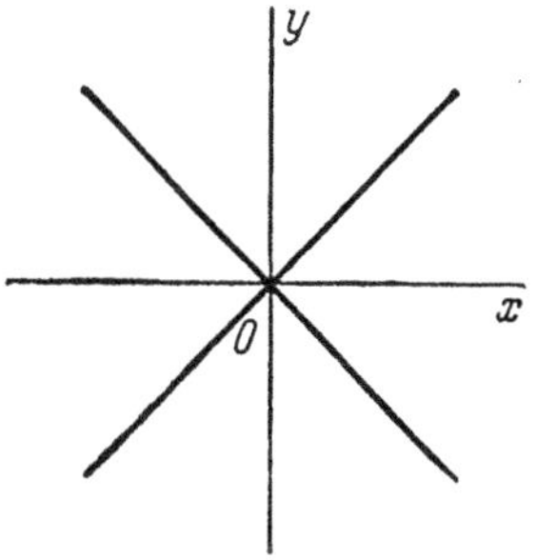

FIG. 60

of the angle between the axes lying in the 1st and 3rd quadrants. 10) The curve consists of 2 straight lines: the x-axis itself, and the bisector of the angle between the axes lying in the 2nd and 4th quadrants. 11) The curve consists of the 2 bisectors of the angle between the coordinate axes (Fig. 60). 12) The curve consists of 2 straight lines — the x- and y-axes. 13) The curve consists of two straight lines parallel to the x-axis which cut off on the y-axis intervals equal to 3 and -3, measured from the origin (Fig. 61). 14) the curve consists of two straight lines parallel to the y-axis which cut off on the x-axis intervals equal 3 and 5, measured from origin (Fig. 62).

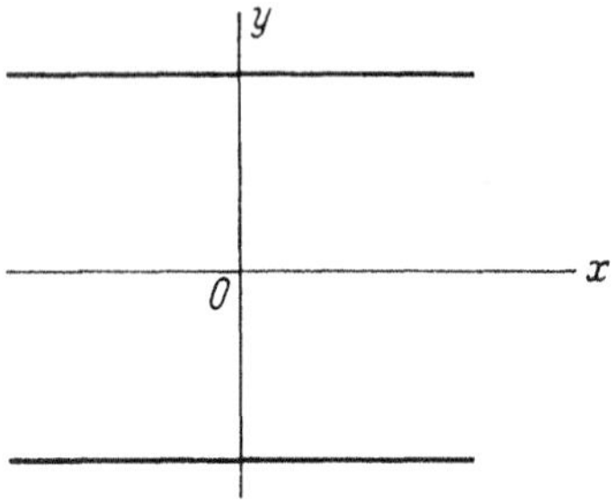

FIG. 61

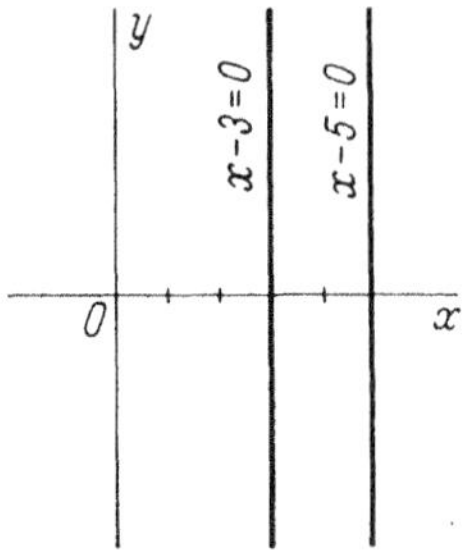

FIG. 62

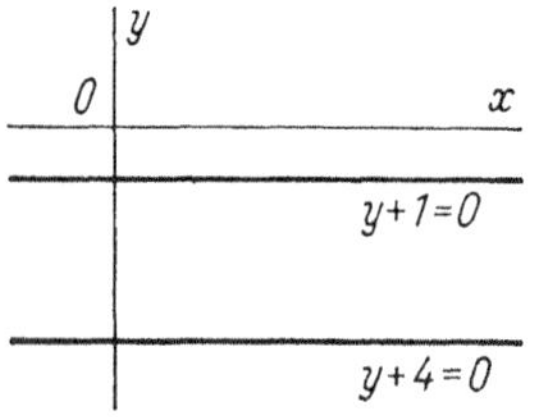

FIG. 63

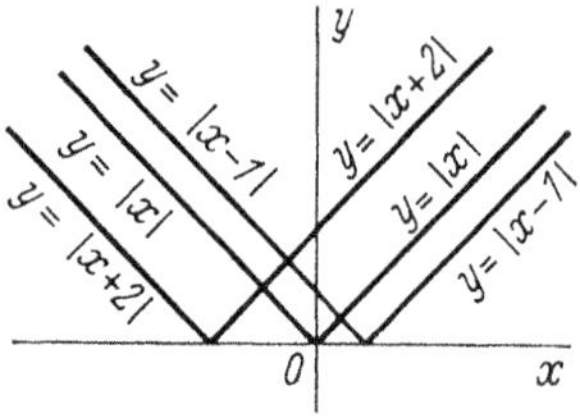

FIG. 64

15) The curve consists of two straight lines parallel to the x-axis which cut off on the negative y-axis intervals equal to 1 and 4 (Fig. 63). 16) The curve consists of three straight lines: one is the x-axis, and the other two, parallel to the y-axis, cut off on the positive x-axis intervals equal to 2 and 5, measured from the origin. 17) The curve consists of two rays: the bisectors of the angles between the axes in the 1st and 2nd quadrants (Fig. 64). 18) The curve consists of two rays: the bisectors of the angles between the axes in the 1st and 4th quadrants [Fig. 65(a)]. 19) The curve consists of two rays: the bisectors of the angles between the axes in the 3rd and 4th quadrants [Fig. 65(b)]. 20) The curve consists of two rays: the bisectors of the angles between the axes in the 2nd and 3rd quadrants [Fig. 65(c)]. 21) The curve consists of two rays in the upper half-plane, coming from the point (1, 0), and parallel to the bisectors of the angles between the axes (Fig. 64). 22) The curve consists of two rays in the upper half-plane, coming from the point $(-2, 0)$, and parallel to

the bisectors of the angles between the axes (Fig. 64). 23) Circle, centre at origin, and radius 4 (Fig. 66). 24) Circle, centre at (2, 1), and radius 4 (Fig. 66). 25) Circle, centre at (−5, 1), and radius 3. 26) Circle, centre at (1, 0), and radius 2. 27) Circle, centre at (0, −3), and radius 1. 28) The curve consists of the single point (3, 0) — a degenerate curve. 29) The

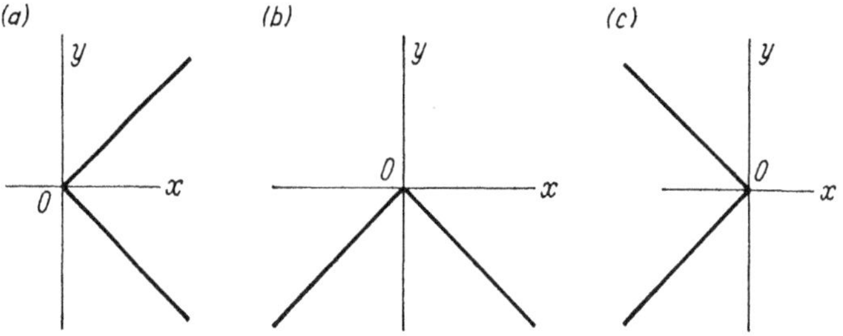

FIG. 65

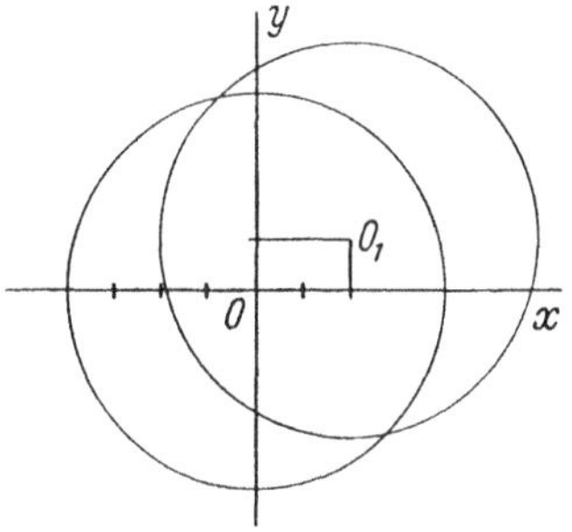

FIG. 66

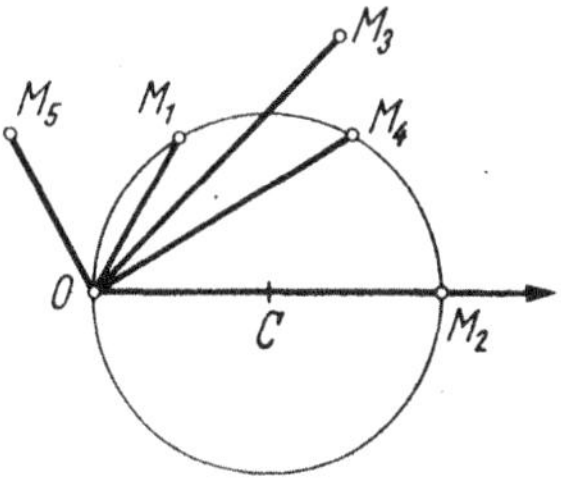

FIG. 67

curve consists of the single point (0, 0) — a degenerate curve. 30) There is no real point whose coordinates satisfy this equation, which defines an "imaginary curve". 31) As for 30). **160.** The curves defined by 1), 2), 4) go through the origin. **161.** 1) a) (7, 0), (−7, 0) b) (0, 7), (0, −7). 2) a) (0, 0), (6, 0), b) (0, 0), (0, −8). 3) a) (−10, 0), (−2, 0, b) curve does not intersect the y-axis. 4) curve does not intersect either axis. 5) a) (0, 0), (12, 0), b) (0, 0), (0, −16). 6) a) curve does not intersect the x-axis, b) (0, −1), (0, −7). 7) curve does not intersect either axis. **162.** 1) (2, 2), (−2, −2). 2) (1, −1), (9, −9). 3) (3, −4), (7/5, −24/5). 4) curves do not intersect. **163.** M_1, M_2, M_4 lie on the curve, which is a circle (Fig. 67). **164.** a) $(6, \pi/3)$, b) $(6, -\pi/3)$, c) (3, 0), d) $(2\sqrt{3}, \pi/6)$. A straight line perpendicular to the polar axis and cutting off on it an interval equal to 3, measured from the pole (Fig. 68). **165.** a) $(1, \pi/2)$, b) $(2, \pi/6)$ and $(2, 5\pi/6)$, c) $(\sqrt{2}, \pi/4)$ and $(\sqrt{2}, 3\pi/4)$. A straight line in the upper half-plane parallel to the polar axis at distance 1 from it (Fig. 68) **166.** 1) Circle, centre at pole, radius 5. 2) A ray starting from the pole and inclined to the polar axis

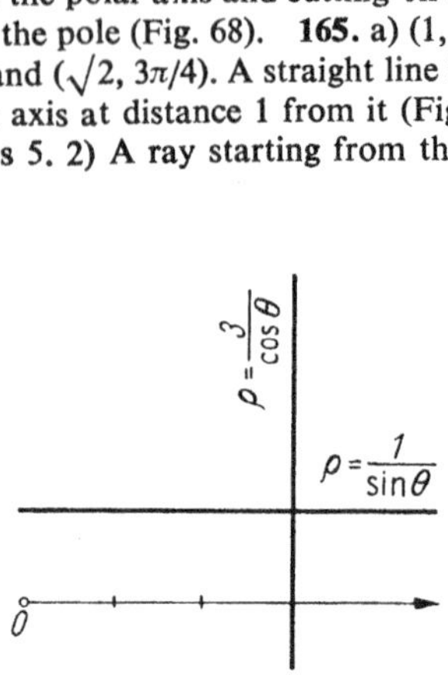

FIG. 68

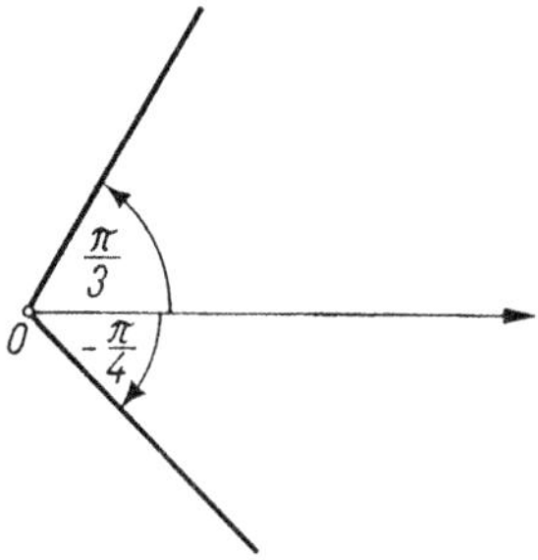

FIG. 69

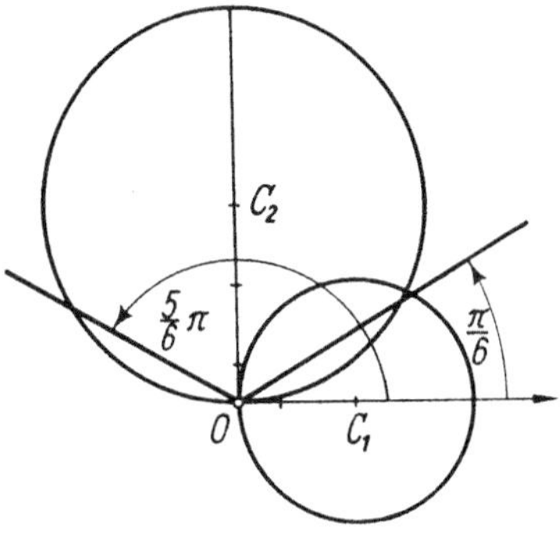

FIG. 70

at an angle of $\pi/3$ (Fig. 69). 3) A ray starting from the pole and inclined to the polar axis at an angle of $-\pi/4$ (Fig. 69). 4) A straight line perpendicular to the polar axis and cutting off on it an interval $a = 2$, measured from the pole. 5) A straight line in the upper half-plane, parallel to the polar axis, and at distance 1 from it. 6) Circle, centre C_1 (3, 0) and radius 3 (Fig. 70). 7) Circle, centre $C_2(5, \pi/2)$, and radius 5 (Fig. 70). 8) The curve consists of two rays, starting at the pole, and inclined to the polar axis at the angles $\pi/6$ and $5\pi/6$ (Fig. 70). 9) The curve consists of a family of concentric circles, centred at the pole, and with radii $r = (-1)^n\pi/6 + n\pi$, where n is any positive integer or zero. **167.** See Figs. 71 and 72. **168.** See Figs. 73 and 74. **169.** See Fig. 75. **170.** See Fig. 76. The small interval attached to the pole has a length $\pi/2$, and each of the other intervals a length 6π. **171.** Into 5 parts (Fig. 77). **172.** $P(12, 1/2)$ (Fig. 78).

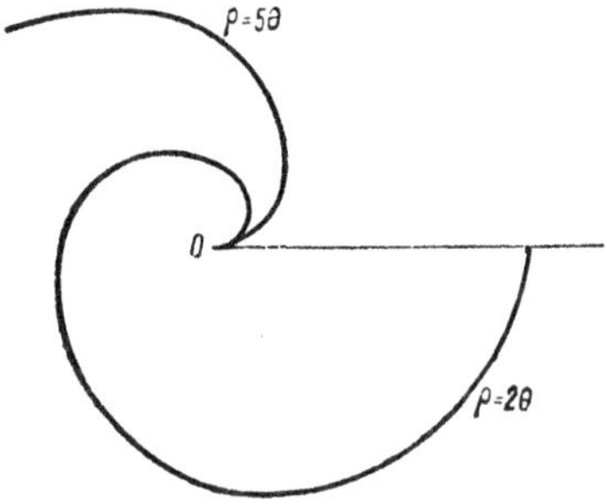

FIG. 71

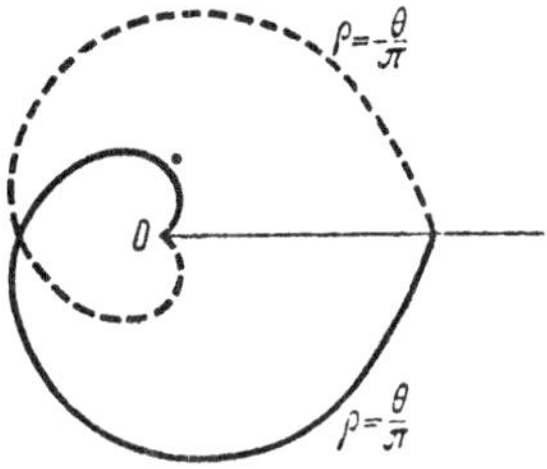

FIG. 72

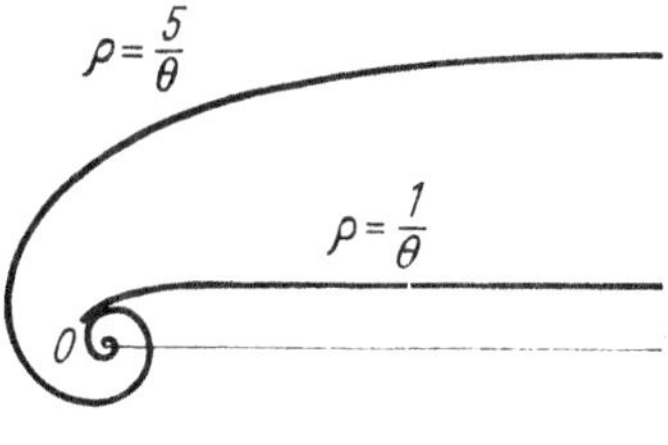

FIG. 73

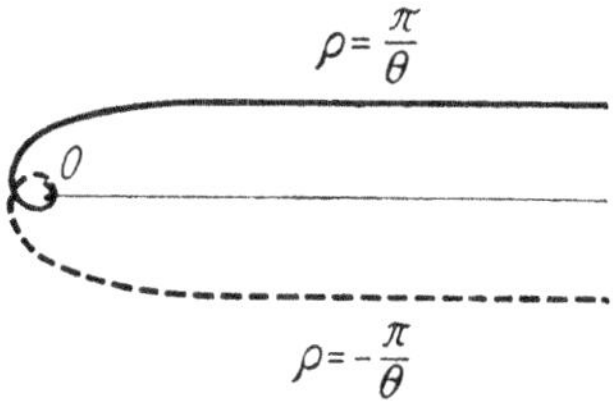

FIG. 74

173. $Q(81, 4)$ (Fig. 79). **174.** $x \pm y = 0$. **175.** $x \pm a = 0$. **176.** $y \pm b = 0$. **177.** $y+4 = 0$. **178.** $x-5 = 0$. **179.** 1) $x-y = 0$, 2) $x+y = 0$, 3) $x-1 = 0$, 4) $y-2 = 0$. **180.** $4ax \pm c = 0$. **181.** $x^2+y^2 = r^2$. **182.** $(x-\alpha)^2+(y-\beta)^2 = r^2$. **183.** $x^2+y^2 = 9$. **184.** $x^2+y^2 = 16$. **185.** $x^2+y^2 = a^2$. **186.** $(x-4)^2+y^2 = 16$. **187.** $x^2/25+y^2/16 = 1$. **188.** $x^2/9-y^2/16 = 1$. **189.** $y^2 = 12x$. **192.** $y^2 = 4ax$. **193.** $x^2/25+y^2/9 = 1$: locus is an ellipse. **194.** $x^2/16-y^2/9 = 1$: locus is a hyperbola. **195.** $x^2/25+6^2/16 = 1$: locus is an ellipse. **196.** The right-hand branch of the hyperbola given by $x^2/64-y^2/36 = 1$. **197.** $y^2 = 20x$: locus is a parabola. **198.** $\varrho \cos \theta = 3$. **199.** $\theta = \pi/3$. **200.** $\tan \theta = 1$. **201.** $\varrho \sin \theta + 5 = 0$, $\varrho \sin \theta - 5 = 0$. **202.** $\varrho = 10 \cos \theta$. **203.** The conditions of the problem are satisfied by two circles, whose polar equations are $\varrho + 6 \sin \theta = 0$, $\varrho - 6 \sin \theta = 0$. **204.** $x = a \cos t$, $y = b \sin t$; $x^2/a^2+y^2/b^2 = 1$. **205.**

$x = (ab\cos t)/D$, $y = (ab\sin t)/D$ where $D = \sqrt{(a^2\sin^2 t + b^2\cos^2 t)}$. **206.** $x = (ab\cos t)/\sqrt{(b^2\cos^2 t - a^2\sin^2 t)}$, $y = (ab\sin t)/\sqrt{(b^2\cos^2 t - a^2\sin^2 t)}$ **207.** 1) $x = t^2/2p$, $y = t$, 2) $x = 2p\cot^2 t$, $y = 2p\cot t$,

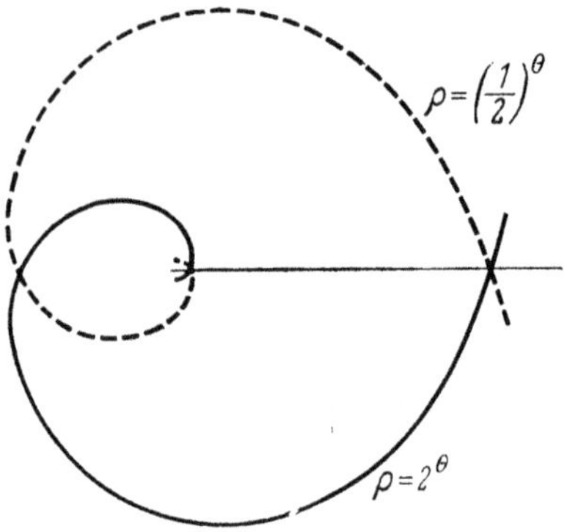

FIG. 75

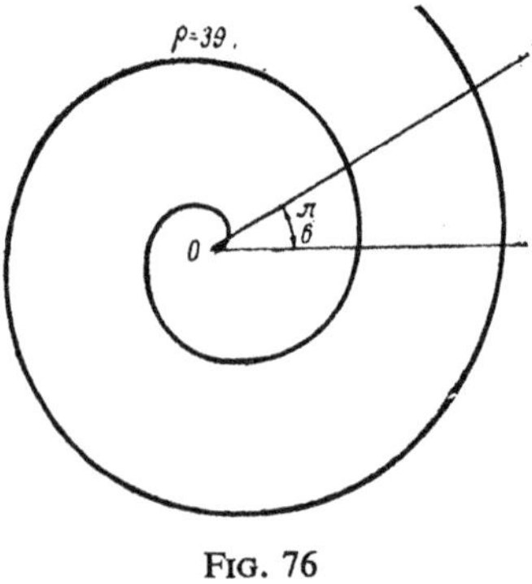

FIG. 76

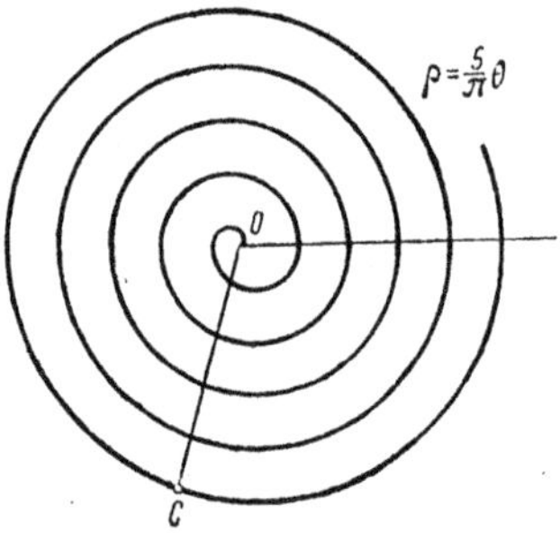

FIG. 77

3) $x = \dfrac{p}{2}\cot^2\dfrac{t}{2}$, $y = p\cot\dfrac{t}{2}$. **208.** 1) $x = 2R\cos^2\theta$, $y = R\sin 2\theta$, 2) $x = R\sin 2\theta$, $y = 2R\sin^2\theta$, 3) $x = 2p\cot^2\theta$, $y = 2p\cot\theta$. **209.** 1) $x - y^2 = 0$, 2) $x^2 + y^2 - a^2 = 0$, 3) $x^2/a^2 - y^2/b^2 - 1 = 0$, 4) $x^2/a^2 - y^2/b^2 - 1 = 0$, 5) $x^2 + y^2 - 2Rx = 0$, 6) $x^2 + y^2 - 2Ry = 0$, 7) $2px - y^2 = 0$. **210.** M_1, M_3, M_4. **211.** 3, −3, 0, −6. **212.** 1, −2, 4, −5, 7. **213.** (6, 0), (0, −4). **214.** (3, −5). **215.** $A(2, -1)$, $B(-1, 3)$, $C(2, 4)$. **216.** (1, −3), (−2, 5), (5, −9), (8, −17). **217.** $S = 17$ units of area.

218. $C_1(-1, 4)$ or $C_2(25/7, -36/7)$. **219.** $C_1(1, -1)$ or $C_2(-2, -10)$ **220.** 1) $2x-3y+9=0$, 2) $3x-y=0$, 3) $y+2=0$, 4) $3x+4y-12=0$, 5) $2x+y+5=0$, 6) $x+3y-2=0$. **221.** 1) $m=5$, $c=3$, 2) $m=-2/3$, $c=2$, 3) $m=-5/3$, $c=-2/3$, 4) $m=-3/2$, $c=0$, 5) $m=0$, $c=3$. **222.** 1) $-5/3$, 2) $3/5$. **223.** 1) $2x+3y-7=0$, 2) $3x-2y-4=0$. **224.** $3x+2y=0$, $2x-3y-13=0$. **225.** (2, 1), (4, 2), (−1 ,7), (1, 8).

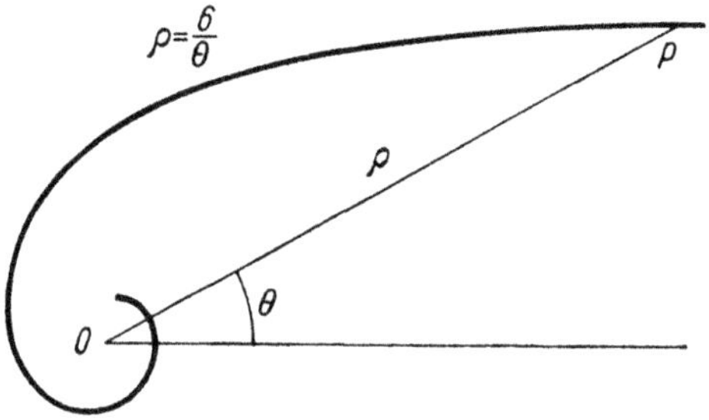

FIG. 78

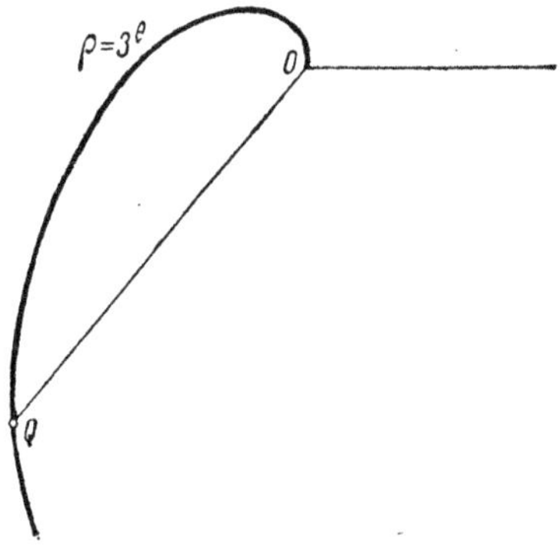

FIG. 79

226. (−2, −1). **227.** $Q(11, -11)$. **228.** 1) $3x-2y-7=0$, 2) $5x+y-7=0$, 3) $8x+12y+5=0$, 4) $5x+7y+9=0$, 5) $6x-30y-7=0$. **229.** *a)* $m=7$, *b)* $m=7/10$, *c)* $m=-3/2$. **230.** $5x-2y-33=0$, $x+4y-11=0$, $7x+6y+33=0$. **231.** $7x-2y-12=0$, $5x+y-28=0$, $2x-3y-18=0$. **232.** $x+y+1=0$. **233.** $2x+3y-13=0$. **234.** $4x+3y-11=0$, $3x+2y-13=0$, $x+y+2=0$. **235.** (3, 4). **236.** $4x+y-3=0$. **237.** $x-5=0$. **238.** Equations for sides: AB $2x+y-8=0$, BC $x+2y-1=0$, CA $x-y-1=0$. Equations for medians, through A $x-3=0$, through B $x+y-3=0$, through C $y=0$. **239.** $(-7, 0), (0, 7/3)$. **242.** $(1, 3)$. **243.** $3x-5y+4=0, x+7y-16=0, 3x-5y-22=0, x+7y+10=0$. **244.** Equations of sides: $2x-5y+3=0$, $2x-5y-26=0$. Equations of diagonals: $7x-3y-33=0$. **245.** Internal bisector: $5x+y-3=0$. External bisector: $x-5y-11=0$. **246.** $x+y-8=0$, $11x-y-28=0$. *Note:* The conditions of the problem are satisfied by two straight lines: one through P and the mid-point of AB, the other through P and parallel to AB. **247.** $(-12, 5)$. **248.** $M_1(10, -5)$. **249.** $P(5/3, 0)$. *Hint:* P is a point such that the inclinations of PM and PN to the x-axis are the same. **250.** $P(0, 11)$. **251.** $P(2, -1)$. **252.** $P(2, 5)$. **253.** 1) $\varphi=\pi/4$, 2) $\varphi=\pi/2$, 3) $\varphi=0$ the lines are parallel, 4) $\varphi=\tan^{-1} 16/11$. **254.** $x-5y+3=0$ or $5x+y-11=0$. **255.** Equations of sides: $4x+3y+1=0$, $3x-4y+32=0$, $4x+3y-24=0$, $3x-4y+7=0$. Equation of diagonal: $x+7y-31=0$. **256.** $3x-4y+15=0$, $4x+3y-30=0$, $3x-4y-10=0$, $4x+3y-5=0$. **257.** $2x+y-16=0$, $2x+y+14=0$, $x-2y-18=0$. **258.** $3x-y+9=0$, $3x+y+9=0$. **259.** $29x-2y+33=0$. **262.** 1) $3x-7y-27=0$, 2) $x+9y+25=0$, 3) $2x-3y-13=0$, 4) $x-2=0$, 5) $y+3=0$. **264.** 1), 3), 4). **266.** 1) $\varphi=45°$, 2) $\varphi=60°$, 3) $\varphi=90°$. **267.** $M_3(6, -6)$. **268.** $4x-y-13=0$, $x-5=0$, $x+8y+5=0$. **269.** BC: $3x+4y-22=0$. CA: $2x-7y-5=0$. CN: $3x+5y-23=0$. **270.** $x+2y-7=0$, $x-4y-1=0$, $x-y+2=0$. *Hint:* Find $G(\bar{x}, \bar{y})$ the point of intersection of the medians. If the vertices are $A(x_1, y_1)$, $B(x_2, y_2)$ $C(x_3, y_3)$, then $x_1+x_2+x_3=3\bar{x}$, $y_1+y_2+y_3=3\bar{y}$. Also, B lies on one of the given medians, C on the other. Hence we have 4 simple equations for the unknowns x_2, x_3, y_2, y_3. Having found the vertices of the triangle, the equations of its sides can be written down. **271.** $3x-5y-13=0$, $8x-3y+17=0$, $5x+2y-1=0$. *Hint:* The sides through A are respectively perpendicular to the given altitudes. **272.** $2x-y+3=0$, $2x+y-7=0$, $x-2y-6=0$. *Hint:* the mirror-images of A in the given internal bisectors each lie on the side opposite to A. **273.** $4x-3y+10=0$, $7x+y-20=0$, $3x+4y-5=0$. **274.** $4x+7y-1=0$, $y-3=0$, $4x+3y-5=0$. **275.** $3x+7y-5=0$, $3x+2y-10=0$, $9x+11y+5=0$. **276.** $x-3y-23=0$, $7x+9y+19=0$, $4x+3y+13=0$. **277.** $x+y-7=0$, $x+7y+5=0$, $x-8y+20=0$. **278.** $2x+9y-65=0$, $6x-7y-25=0$, $18x+13y-41=0$. **279.** $x+2y=0$, $23x+25y=0$. **280.** $8x-y-24=0$. **283.** $3x+y=0$, $x-3y=0$. **284.** $3x+4y-1=0$, $7x+24y-61=0$. **285.** 1) $a=-2$, $5y-33=0$, 2) $a_1=-3$, $x-56=0$, $a_2=3$, $5x+8=0$, 3) $a_1=1$, $3x-8y=0$, $a_2=5/3$, $33x-56y=0$. **286.** $m=7$, $n=-2$, $y+3=0$. **287.** $m=-4$, $n=2$, $x-5=0$. **288.** 1) $(5, 6)$, 2) $(3, 2)$, 3) $(1/4, 1/3)$, 4) $(2, -1/11)$, 5) $(-5/3, 2)$. **291.** 1) For $a\neq 3$, 2) For $a=3$ and $b\neq 2$, 3) For $a=3$ and $b=2$. **292.** 1) $m=-4$, $n\neq 2$ or $m=4$, $n\neq -2$, 2) $m=-4$, $n=2$ or $m=4$, $n=-2$, 3) $m=0$, and n arbitrary. **293.** $m=7/12$. **294.** The conditions of the problem are satisfied by two values of m: $m_1=0$, $m_2=6$.

295. 1) Yes, 2) No, 3) No. **298.** $a = -7$. **299.** 1) $x/3+y/2 = 1$, 2) $x/-6+y/8 = 1$, 3) $x/(\frac{9}{2})+y/3 = 1$, 4) $x/(\frac{2}{3})+y/(-\frac{2}{5}) = 1$, 5) $x/(\frac{1}{5})+ +y/(\frac{1}{2}) = 1$ (see Fig. 80). **300.** 6 units of area. **301.** $x+y+4 = 0$. **302.** $x+y-5 = 0$, $x-y+1 = 0$, $3x-2y = 0$. **303.** *Solution.* The straight line given by $x/a+y/b = 1$ goes through $C(1, 1)$ if

$$1/a+1/b = 1 \qquad (1)$$

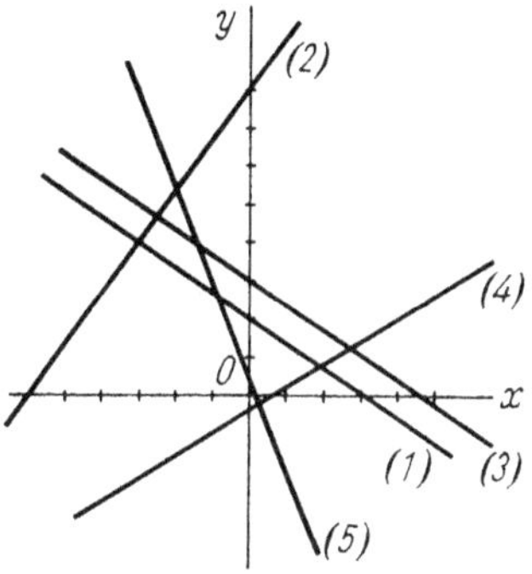

FIG. 80

and it cuts off intervals a, b on the axes. The area S of the triangle enclosed between the straight line and the axes is given by

$$S = \pm ab/2 \qquad (2)$$

the sign being + if a, b have the same sign, and − if they have different signs. Here $S = 2$ and from (2), (1) we have

$$ab = \pm 2S = \pm 4 \quad \text{and} \quad a+b = \pm 4.$$

Hence a, b are the roots of one or other of the quadratic equations

$$p^2-4p+4 = 0 \quad \text{or} \quad p^2+4p-4 = 0.$$

Therefore $$p = 2 \quad \text{or} \quad p = -2\pm 2\sqrt{2}.$$

Thus the conditions of the problem are satisfied by each of the three straight lines given by

$$\frac{x}{2}+\frac{y}{2} = 1 \quad \text{or} \quad \frac{x}{-2+2\sqrt{2}}+\frac{y}{-2-2\sqrt{2}} = 1$$

$$\text{or} \quad \frac{x}{-2-2\sqrt{2}}+\frac{y}{-2+2\sqrt{2}} = 1$$

i.e., by $x+y-2 = 0$, or $(\sqrt{2}+1)x-(\sqrt{2}-1)y-2 = 0$, or $(\sqrt{2}+1)y-(\sqrt{2}-1)x-2 = 0$.

304. The conditions of the problem are satisfied by the three straight lines given by $(\sqrt{2}+1)x+(\sqrt{2}-1)y-10=0$,
$(\sqrt{2}+1)y+(\sqrt{2}-1)x+10=0$, and $x-y-10=0$.
305. $3x-2y-12=0$, $3x-8y+24=0$. **306.** $x+3y-30=0$. $3x+4y-60=0$, $3x-y-30=0$, $x-12y+60=0$. **307.** The conditions of the problem are satisfied by two straight lines which intersect the axes in the points (2, 0), (0, −3) and (−4, 0), (0, 3/2). **308.** $S \geq 2x_1y_1$. **309.** 1), 4), 6), 8). **310.** 1) $4x/5-3y/5-2=0$, 2) $-4x/5+3y/5-10=0$, 3) $-12x/13+5y/13-1=0$, 4) $-x-2=0$, 5) $2x/\sqrt{5}-y/\sqrt{5}-1=0$. **311.** 1) $\alpha=0$, $p=2$, 2) $\alpha=\pi$, $p=2$, 3) $\alpha=\pi/2$, $p=3$, 4) $\alpha=-\pi/2$, $p=3$, 5) $\alpha=\pi/6$, $p=3$, 6) $\alpha=-\pi/4$, $p=\sqrt{2}$, 7) $\alpha=-2\pi/3$, $p=1$, 8) $\alpha=-\beta$, $p=g$, 9) $\alpha=\beta-\pi$, $p=g$. **312.** 1) $\delta=-3$, $d=3$, 2) $\delta=1$, $d=1$, 3) $\delta=-4$, $d=4$, 4) $\delta=d=0$ and Q lies on the straight line. **313.** On same side, 1, 3, 4: on opposite sides 2,5. **314.** 5 units of area. **315.** 6 units of area. **318.** Convex. **319.** Not convex. **320.** 4. **321.** 3. **322.** 1) $d=2{\cdot}5$, 2) $d=3$, 3) $d=0{\cdot}5$, 4) $d=3{\cdot}5$. **323.** 49 units of area. **325.** In the ratio 2:3, measured from the second line. **326.** *Solution:* The problem is equivalent to that of drawing tangents from the point $P(2, 7)$ to the circle, centre $Q(1, 2)$ and radius 5. Since $PQ=\sqrt{[(2-1)^2+(7-2)^2]}=\sqrt{26}$, and this is greater than 5, the point P is outside the circle and there are two such real tangents. Next we have to find their equations. Any line through P is given by an equation of the form

$$y-7=m(x-2), \quad \text{or} \quad mx-y+7-2m=0, \tag{1}$$

where m is the gradient to be found. Equation (1) is put into normal form by multiplying by the normalizing factor $\mu=\pm 1/\sqrt{(m^2+1)}$: we then get

$$\frac{mx-y+7-2m}{\pm(m^2+1)}=0. \tag{2}$$

Substituting the coordinates of Q we have $\dfrac{|m-2+7-2m|}{\sqrt{(m^2+1)}}=5$,

giving $m_1=-5/12$, $m_2=0$. Hence the equations of the required lines are $y-7=-5(x-2)/12$, or $5x+12y-94=0$, and $y-7=0$. **327.** $7x+24y-134=0$, $x-2=0$. **328.** $3x+4y-13=0$. **330.** $8x-15y+9=0$. **331.** $3x-4y-25=0$, $3x-4y+5=0$. **332.** The conditions of the problem are satisfied by two squares, symmetrical about AB. The equations of the sides of the one square are: $4x+3y-8=0$, $4x+3y+17=0$, $3x-4y-6=0$, $3x-4y+19=0$; the equations of the sides of the other square are: $4x+3y-8=0$, $4x+3y-33=0$, $3x-4y-6=0$, $3x-4y+19=0$. **333.** There are two squares which satisfy the conditions of the problem. The remaining sides of the first square are given by $3x+4y-11=0$, $4x-3y-23=0$, $3x+4y-27=0$; the remaining sides of the second square are given by $3x+4y-11=0$, $4x-3y-23=0$, $3x+4y+5=0$. **334.** $3x+4y+6=0$, $3x+4y-14=0$ or $3x+4y+6=$

$= 0$, $3x+4y+26 = 0$. **335.** $12x-5y+61 = 0$, $12x-5y+22 = 0$ or $12x-5y+61 = 0$, $12x-5y+100 = 0$. **336.** $M(2, 3)$. **337.** $4x+y+5 = 0$, $y-3 = 0$. **338.** 1) $3x-y+2 = 0$, 2) $x-2y+5 = 0$, 3) $20x-8y-9 = 0$. **339.** 1) $4x-4y+3 = 0$, $2x+2y-7 = 0$, 2) $4x+1 = 0$, $8y+13 = 0$, 3) $14x-8y-3 = 0$, $64x+112y-23 = 0$. **340.** $x-3y-5 = 0$, $3x+y-5 = 0$. *Hint:* the required lines go through P and are perpendicular to the bisectors of the angles between the given lines. **341.** 1) In the same angle, 2) In adjacent angles, 3) In vertically opposite angles. **342.** 1) In vertically opposite angles, 2) In adjacent angles, 3) In the same angle. **343.** Inside the triangle. **344.** Outside the triangle. **345.** The acute angle. **346.** The obtuse angle. **347.** $8x+4y-5 = 0$. **348.** $x+3y-2 = 0$. **349.** $3x-19 = 0$. **350.** $10x-10y-3 = 0$. **351.** $7x+56y-40 = 0$. **352.** $x+y+5 = 0$. **353.** $S(2, -1)$. **354.** 1) $3x+2y-7 = 0$, 2) $2x-y = 0$, 3) $y-2 = 0$, 4) $x-1 = 0$, 5) $4x+3y-10 = 0$, 6) $3x-2y+1 = 0$. **355.** $74x+13y+39 = 0$. **356.** $x-y-7 = 0$. **357.** $7x+19y-2 = 0$. **358.** $x-y+1 = 0$. **359.** $4x-5y+22 = 0$, $4x+y-18 = 0$, $2x-y+1 = 0$. **360.** $x-5y+13 = 0$, $5x+y+13 = 0$. **361.** BC $5x-y-5 = 0$, AC $x-y+3 = 0$, CN $3x-y-1 = 0$. **362.** $x-5y-7 = 0$, $5x+y+17 = 0$, $10x+7y-13 = 0$. **363.** $2x+y+8 = 0$, $x+2y+1 = 0$. **366.** $C = -29$. **367** $a \neq -2$. **368.** Equations of sides of square: $4x+3y-14 = 0$, $3x-4y+27 = 0$, $3x-4y+2 = 0$, $4x+3y+11 = 0$. Equation of second diagonal: $7x-y+13 = 0$. **369.** $x+y+5 = 0$. **370.** $x+y+2 = 0$, $x-y-4 = 0$, $3x+y = 0$. **371.** $2x+y-6 = 0$, $9x+2y+18 = 0$. **372.** $3x-y+1 = 0$. **374.** $3x-4y+20 = 0$, $4x+3y-15 = 0$. **375.** $x+5y-13 = 0$, $5x-y+13 = 0$. **376.** Two straight lines satisfy the conditions of the problem: their equations are $7x+y-9 = 0$, $2x+y+1 = 0$. **377.** $5x-2y-7 = 0$. **378.** AC $3x+8y-7 = 0$, BD $8x-3y+7 = 0$. **379.** $4x+y+5 = 0$, $x-2y-1 = 0$, $2x+5y-11 = 0$. **381.** 1) $\varrho \sin (\beta-\theta) = p$, $\varrho \sin (\pi/6-\theta) = 3$, 2) $\varrho \cos (\theta-\alpha) = a \cos \alpha$, $\varrho \cos (\theta+2\pi/3) = -1$, 3) $\varrho \sin (\beta-\theta) = a \sin \beta$, $\varrho \sin (\pi/6-\theta) = 3$. **382.** $\varrho \sin (\beta-\theta) = \varrho_1 \sin (\beta-\theta_1)$. **383.** $\varrho \cos (\theta-\alpha) = \varrho_1 \cos (\theta_1-\alpha)$. **384.** $\dfrac{\varrho \sin (\theta-\theta_1)}{\varrho_2 \sin (\theta_2-\theta_1)} = \dfrac{\sqrt{\varrho^2+\varrho_1{}^2-2\varrho\varrho_1 \cos (\theta-\theta_1)}}{\sqrt{\varrho_2{}^2+\varrho_1{}^2-2\varrho_2\varrho_1 \cos (\theta_2-\theta_1)}}$

385. 1) $x^2+y^2 = 9$, 2) $(x-2)^2+(y+3)^2 = 49$, 3) $(x-6)^2+(y+8)^2 = 100$, 4) $(x+1)^2+(y-2)^2 = 25$, 5) $(x-1)^2+(y-4)^2 = 8$, 6) $x^2+y^2 = 16$ 7) $(x-1)^2+(y+1)^2 = 4$, 8) $(x-2)^2+(y-4)^2 = 10$, 9) $(x-1)^2+y^2 = 1$, 10) $(x-2)^2+(y-1)^2 = 25$. **386.** $(x-3)^2+(y+1)^2 = 38$. **387.** $(x-4)^2+(y+1)^2 = 5$ and $(x-2)^2+(y-3)^2 = 5$. **388.** $(x+2)^2+(y+1)^2 = 20$. **389.** $(x-5)^2+(y+2)^2 = 20$ and $(x-9/5)^2+(y-22/5)^2 = 20$. **390.** $(x-1)^2+(y+2)^2 = 16$. **391.** $(x+6)^2+(y-3)^2 = 50$ and $(x-29)^2+(y+2)^2 = 800$. **392.** $(x-2)^2+(y-1)^2 = 5$ and $(x-22/5)^2+(y+31/5)^2 = 289/5$. **393.** $(x-2)^2+(y-1)^2 = 81/13$, $(x+8)^2+(y+7)^2 = 25/13$. **394.** $(x-2)^2+(y-1)^2 = 25$ and $(x+202/49)^2+(y-349/49)^2 = (185/49)^2$. **395.** $(x+10/7)^2+(y+25/7)^2 = 1$ and $(x-30/7)^2+(y-5/7)^2 = 1$. **396.** $(x-5)^2+y^2 = 16$, $(x+15)^2+y^2 = 256$, $(x-35/3)^2+(y-40/3)^2 = (32/3)^2$, and $(x-35/3)^2+(y+40/3)^2=(32/3)^2$. **397.** Equations 1), 2), 4), 5), 8), 10) define circles. 1) $C(5, -2)$, $R=5$, 2) C, $(-2, 0)$, $R=8$, 3) equation defines the single point $(5, -2)$, 4) $C\,(0, 5)$, $R=\sqrt{5}$, 5) $C(1, -2)$, $R=5$, 6) the equation does not have any geometrical representation in the plane, 7) the equation defines the single point $(-2, 1)$, 8) $C(-1/2, 0)$, $R=1/2$,

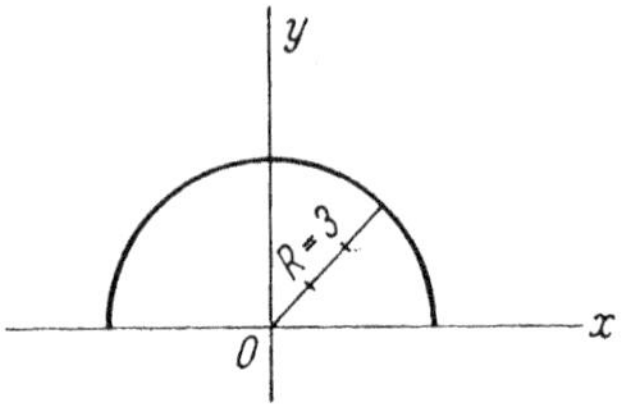

FIG. 81

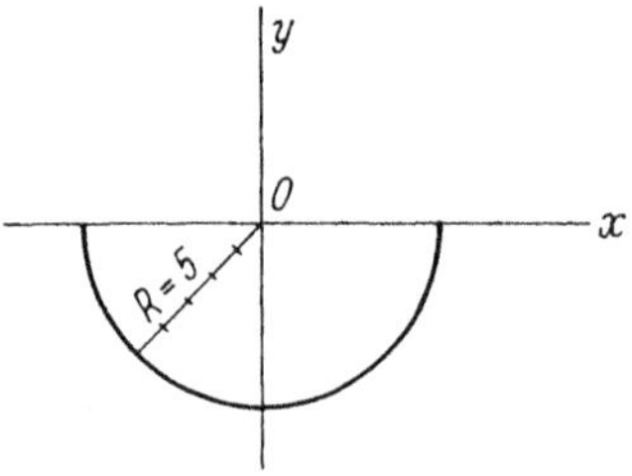

FIG. 82

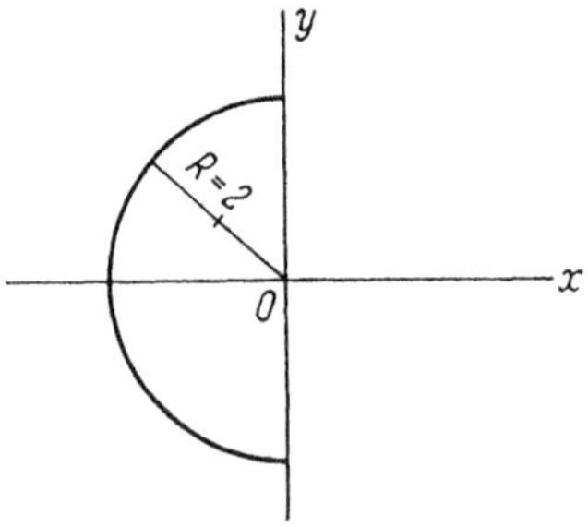

FIG. 83

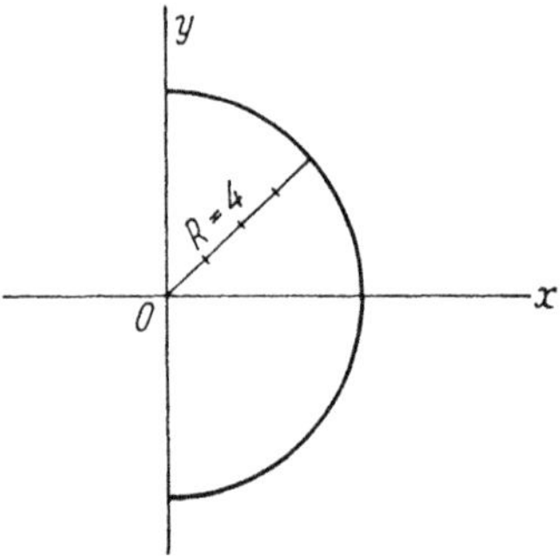

FIG. 84

9) the equation has no geometrical representation in the plane, 10) $C(0, -1/2)$, $R=1/2$. **398.** 1) Semi-circle, radius $R=3$, centre at origin, in upper half-plane (Fig. 81). 2) Semi-circle, radius $R=5$, centre at origin, in lower half-plane (Fig. 82). 3) Semi-circle, radius $R=2$, centre at origin, in left half-plane (Fig. 83). 4) Semi-circle, radius $R=4$, centre at origin, in right half-plane (Fig. 84). 5) Semi-circle, radius $R=8$, centre $C(0, 15)$, situated above the straight line defined by $y-15=0$ (Fig. 85). 6) Semi-circle, radius $R=8$, centre $C(0, 15)$, situated below the straight line defined by $y-15=0$ (Fig. 86). 7) Semi-circle, radius $R=3$, centre $C(-2, 0)$, situated to the left of the straight line defined by $x+2=0$ (Fig. 87). 8) Semi-circle, radius $R=3$, centre $C(-2, 0)$, situated to the right of the straight line defined by $x+2=0$ (Fig. 88). 9) Semicircle, radius $R=5$, centre $C(-2, -3)$, situated below the straight line

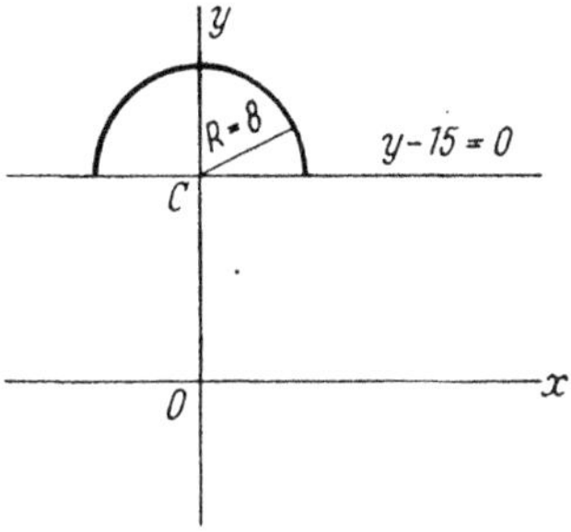

FIG. 85

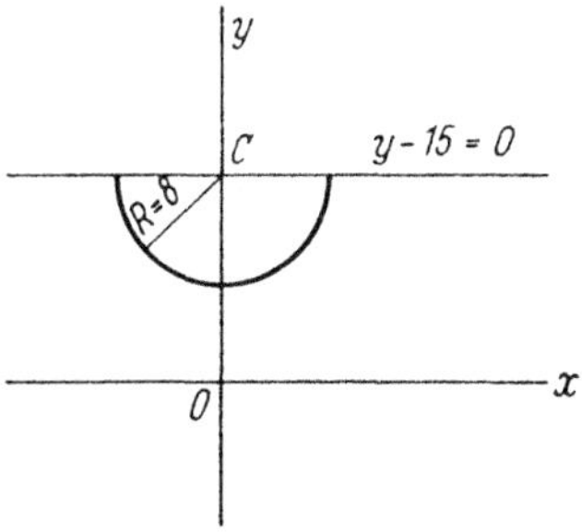

FIG. 86

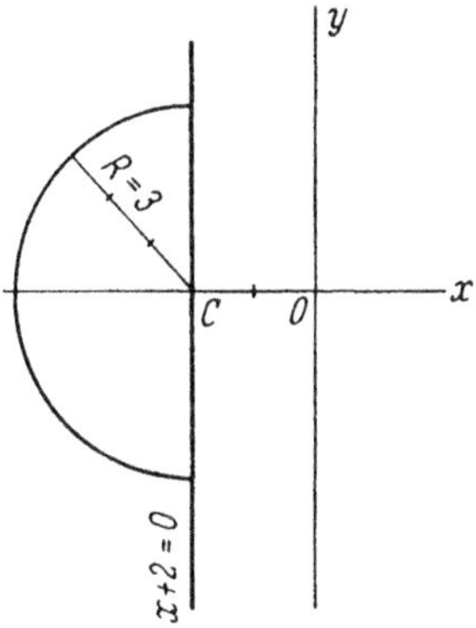

FIG. 87

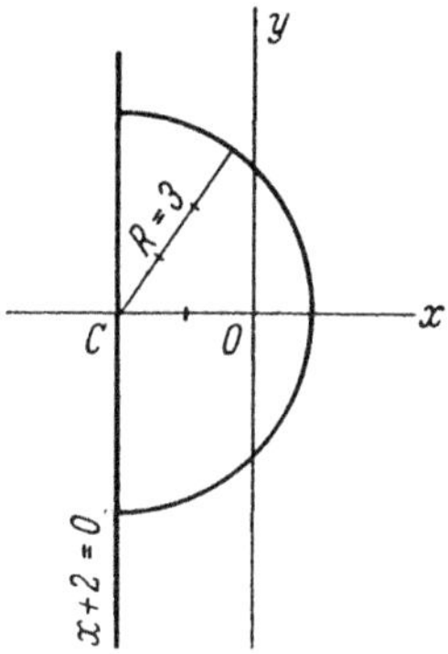

FIG. 88

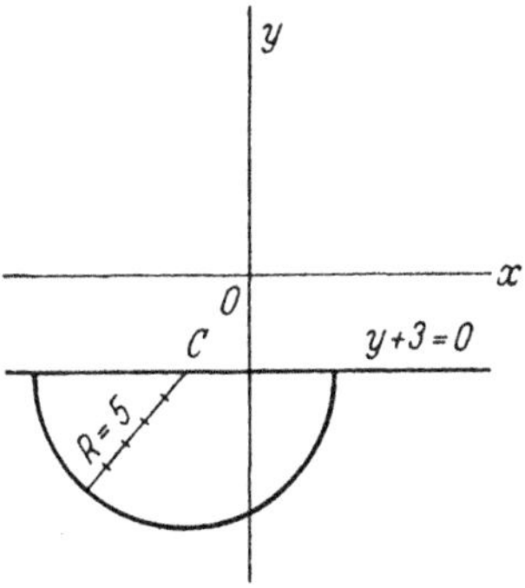

FIG. 89

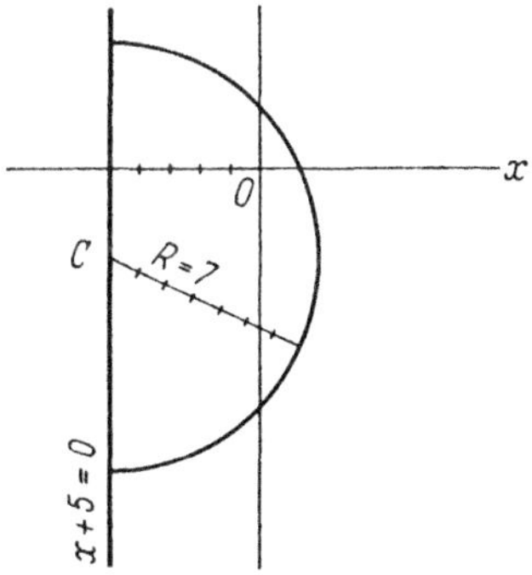

FIG. 90

defined by $y+3=0$ (Fig. 89). 10) Semi-circle, radius $R=7$, centre $C(-5, -3)$, situated to the right of the straight line defined by $x+5=0$ (Fig. 90). **399.** 1) Outside, 2) On circumference, 3) Inside, 4) On circumference, 5) Inside. **400.** 1) $x+5y-3=0$, 2) $x+2=0$, 3) $3x-y-9=0$, 4) $y+1=0$. **401.** $2x-5y+19=0$. **402.** *a*) 7, *b*) 17, *c*) 2. **403.** $M_1(-1, 5)$, $M_2(-2, -2)$. **404.** 1) Intersects, 2) Touches, 3) Passes outside. **405.** 1) $|m|<3/4$, 2) $m=\pm 3/4$, $|m|>3/4$. **406.** $c^2/(1+m^2)=R^2$. **407.** $2x+y-3=0$. **408.** $11x-7y-69=0$. **409.**

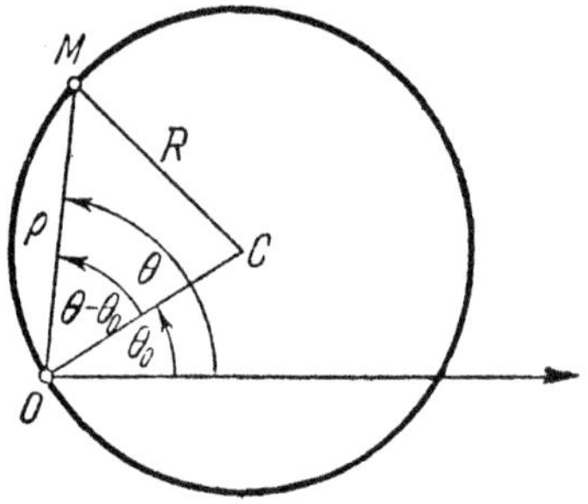

FIG. 91

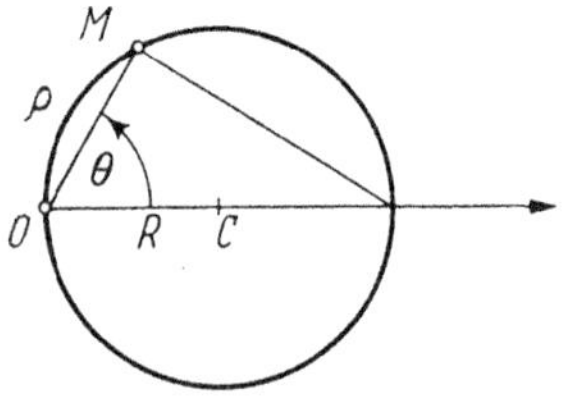

FIG. 92

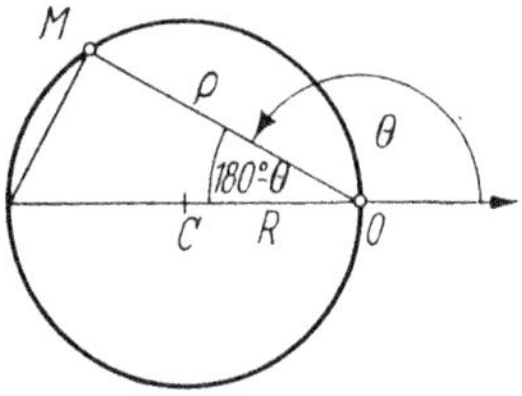

FIG. 93

$2\sqrt{5}$. **410.** $2x-3y+8=0$, $3x+2y-14=0$. **412.** $x^2+y^2+6x-9y-17=0$. **413.** $13x^2+13y^2+3x+71y=0$. **414.** $7x-4y=0$. **415.** 2. **416.** 10. **417.** $(x+3)^2+(y-3)^2=10$. **418.** $x-2y+5=0$. **419.** $3x-4y+43=0$. **420.** $M_1(-7/2, 5/4)$, $d=2\sqrt{5}$. **421.** $x_1x+y_1y=R^2$. **422.** $(x_1-\alpha)(x-\alpha)+(y_1-\beta)(y-\beta)=R^2$. **423.** 45°. **424.** 90°. **425.** $(\alpha_1-\alpha_2)^2+(\beta_1-\beta_2)^2=R_1^2+R_2^2$. **427.** $x-2y-5=0$, $2x-y-5=0$. **428.** $2x+y-8=0$, $x-2y+11=0$. **429.** $2x+y-5=0$, $x-2y=0$. **430.** 90°. **431.** $x+2y+5=0$. **432.** $d=7{\cdot}5$. **433.** $d=6$. **434.** $d=\sqrt{10}$. **435.** 3. **436.** $2x+y-1=0$ and $2x+y+19=0$. **437.** $2x+y-5=0$, $2x+y+5=0$. **438.** $\varrho=2R\cos(\theta-\theta_0)$ (Fig. 91). **439.** 1) $\varrho=2R\cos\theta$ (Fig. 92), 2) $\varrho=-2R\cos\theta$ (Fig. 93), 3) $\varrho=2R\sin\theta$ (Fig. 94),

4) $\varrho = -2R \sin \theta$ (Fig. 95). **440.** 1) (2, 0), $R = 2$, 2) $(3/2, \pi/2)$, $R = 3/2$, 3) $(1, \pi)$, $R = 1$, 4) $(5/2, -\pi/2)$, $R = 5/2$, 5) $(3, \pi/3)$, $R = 3$, 6) $(4, 5\pi/6)$, $R = 4$, 7) $(4, -\pi/6)$, $R = 4$. **441.** 1) $x^2+y^2-3x = 0$, 2) $x^2+y^2+4y = 0$, 3) $x^2+y^2-x+y = 0$. **442.** 1) $\varrho = \cos\theta$, 2) $\varrho = -3\cos\theta$, 3) $\varrho = 5\sin\theta$, 4) $\varrho = -\sin\theta$, 5) $\varrho = \cos\theta+\sin\theta$. **443.** $\varrho = R\sec(\theta-\theta_0)$. **444.** 1) $x^2/25+y^2/4 = 1$, 2) $x^2/25+y^2/9 = 1$, 3) $x^2/169+y^2/144 = 1$, 4) $x^2/25+$ $+y^2/16 = 1$, 5) $x^2/100+y^2/64 = 1$, 6) $x^2/169+y^2/25 = 1$, 7) $x^2/5+y^2 = 1$, 8) $x^2/16+y^2/12 = 1$, 9) $x^2/13+y^2/9 = 1$ or $x^2/(\frac{117}{4})+y^2/9 = 1$, 10) $x^2/64+$

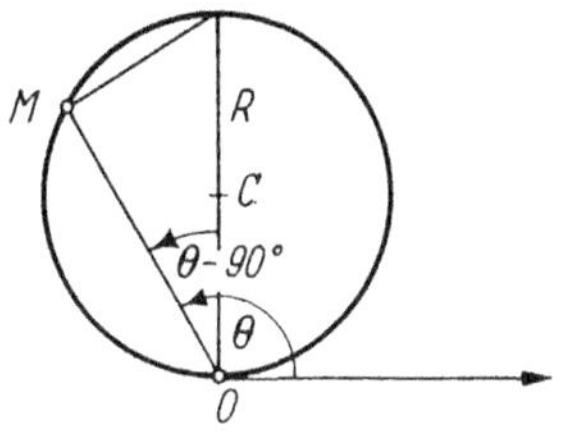

FIG. 94

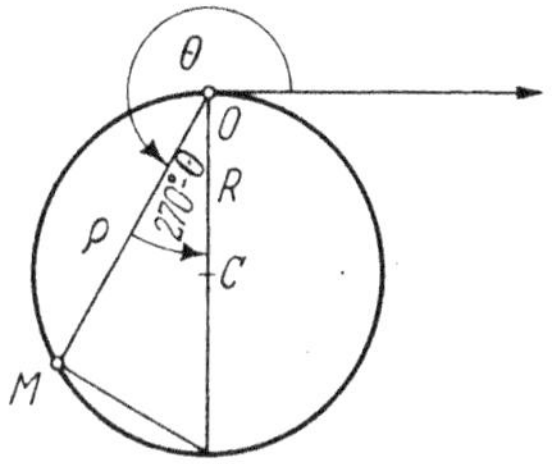

FIG. 95

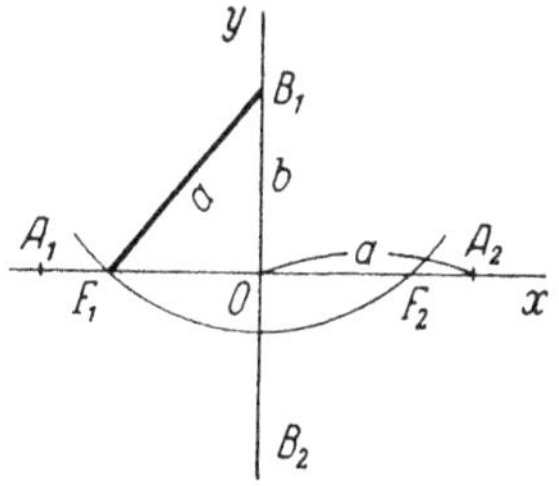

FIG. 96

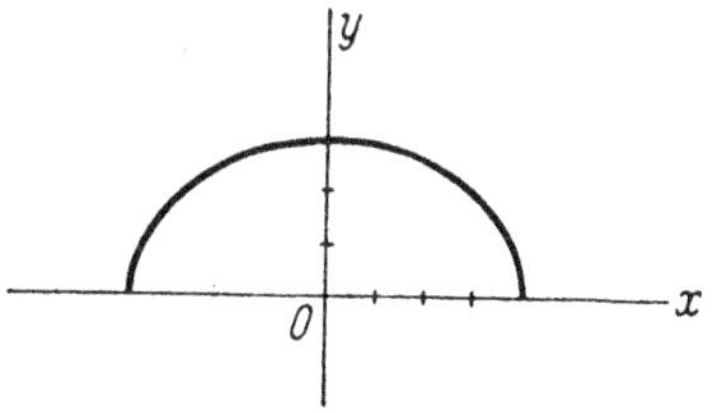

FIG. 97

$+y^2/48 = 1$. **445.** 1) $x^2/4+y^2/49 = 1$, 2) $x^2/9+y^2/25 = 1$, 3) $x^2/25+$ $+y^2/169 = 1$, 4) $x^2/64+y^2/100 = 1$, 5) $x^2/16+y^2/25 = 1$, 6) $x^2/7+y^2/16 =$ $= 1$. **446.** 1) 4 and 3, 2) 2 and 1, 3) 5 and 1, 4) $\sqrt{15}$ and $\sqrt{3}$, 5) 5/2 and 5/3, 6) 1/3 and 1/5, 7) 1 and 1/2, 8) 1 and 4, 9) 1/5 and 1/3, 10) 1/3 and 1. **447.** 1) 5 and 3, 2) $F_1(-4, 0)$, $F_2(4, 0,)$ 3) $\varepsilon = 4/5$, 4) $x = \pm 25/4$. **448.** 16 units of area. **449.** 1) $\sqrt{5}$ and 3, 2) $F_1(0, -2)$, $F_2(0, 2)$, 3) $\varepsilon = 2/3$, 4) $y =$ $= \pm 9/2$. **450.** $4\sqrt{5}/45$ units of area. **451.** b^2/c. **452.** See Fig. 96. **453.** $(-3, -8/5)$, $(-3, 8/5)$. **454.** The points A_1, A_6 lie on the ellipse, A_2, A_4, A_8 inside, and A_3, A_5, A_7, A_9, A_{10} outside the ellipse. **455.** 1) The half of the ellipse given by $x^2/16+y^2/9 = 1$ lying in the upper half-plane (Fig. 97). 2) the half of the ellipse given by $x^2/9+y^2/25 = 1$ lying in the lower half-plane (Fig. 98). 3) the half of the ellipse given by $x^2/4+y^2/9 = 1$

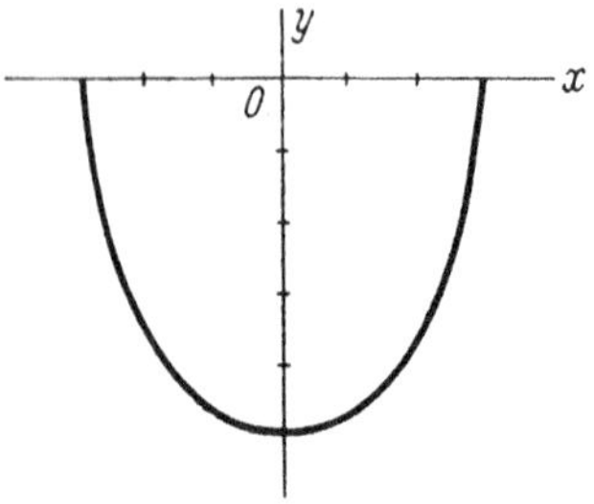

FIG. 98

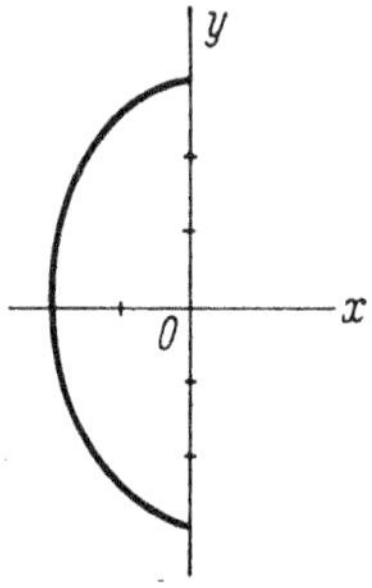

FIG. 99

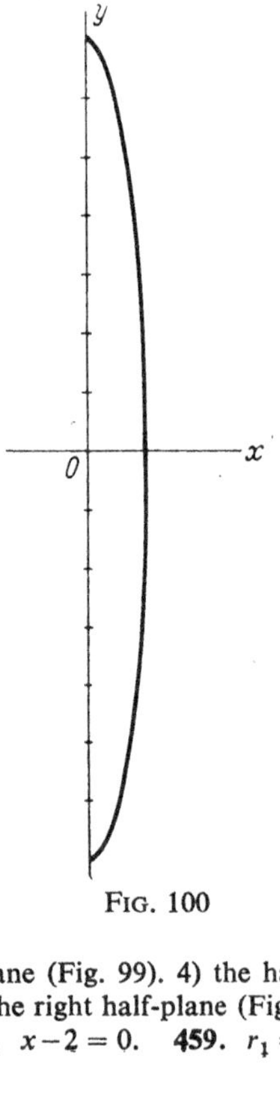

FIG. 100

lying in the left half-plane (Fig. 99). 4) the half of the ellipse given by $x^2+y^2/49=1$ lying in the right half-plane (Fig. 100). **456.** 15. **457.** 8. **458.** $5x+12y+10=0$, $x-2=0$. **459.** $r_1=2{\cdot}6$, $r_2=7{\cdot}4$, **460.** 20,

461. 10. **462.** $(-5, 3\sqrt{3})$, $(-5, -3\sqrt{3})$. **463.** $(-2, \sqrt{21}/21)$, $(-2, -\sqrt{21}/2)$. **464.** 3 and 7. **465.** 1) $x^2/36+y^2/9=1$, 2) $x^2/16+y^2/16=1$, 3) $x^2/20+y^2/15=1$, 4) $x^2/20+y^2/4=1$, 5) $x^2/9+y^2/5=1$, 6) $x^2/256+ +y^2/192=1$, 7) $x^2/15+y^2/6=1$. **466.** 1) $\sqrt{3}/2$, 2) $\sqrt{2}/2$, 3) $\sqrt{3}/3$, 4) 1/2. **467.** $\varepsilon=\sqrt{2}/2$. **468.** $(x-x_0)^2/a^2+(y-y_0)^2/b^2=1$. **469.** $(x- -3)^2/9+(y+4)^2/16=1$. **470.** $(x+3)^2/9+(y-2)^2/4=1$. **471.** 1) $C(3, -1)$, semi-axes 3 and $\sqrt{5}$, $\varepsilon=2/3$, equations of directrices $2x-15=0$,

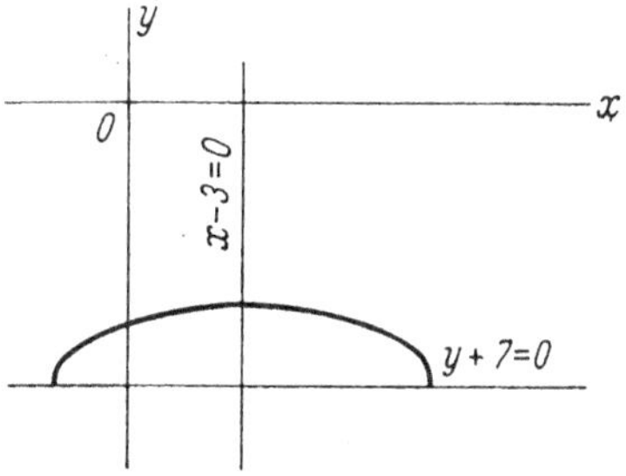

FIG. 101

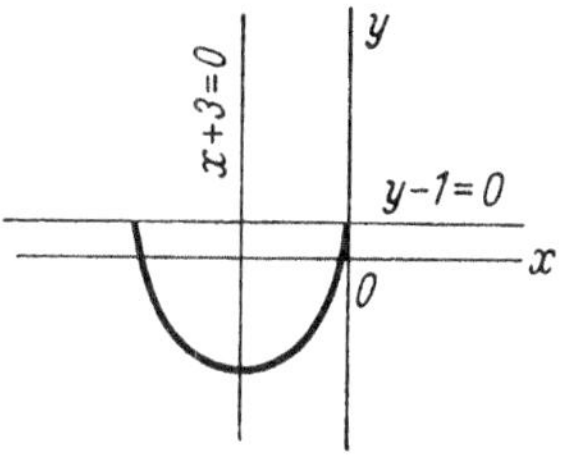

FIG. 102

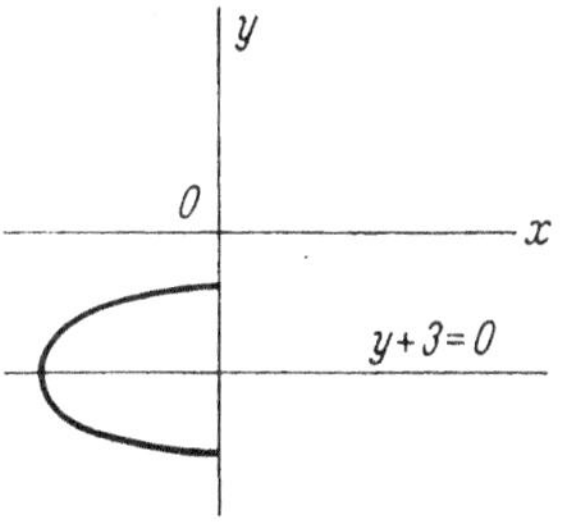

FIG. 103

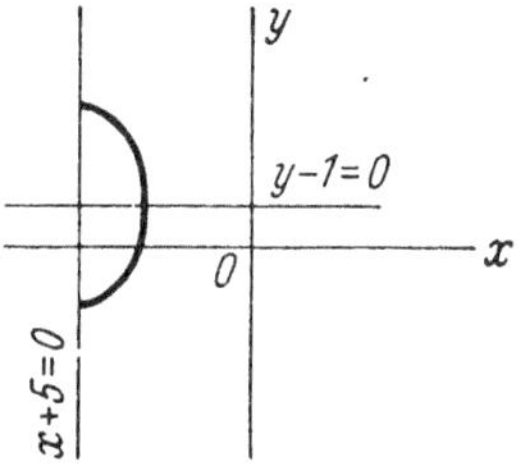

FIG. 104

$2x+3=0$. 2) $C(-1, 2)$, semi-axes 5 and 4, $\varepsilon=3/5$, equations of directrices $3x-22=0$, $3x+28=0$. 3) $C(1, -2)$, semi-axes $2\sqrt{3}$ and 4, $\varepsilon= =1/2$, equations of directrices $y-6=0$, $y+10=0$. **472.** 1) The half of the ellipse given by $(x-3)^2/25+(y+7)^2/4=1$ situated above the straight line given by $y+7=0$ (Fig. 101). 2) The half of the ellipse given by $(x+3)^2/9+(y-1)^2/16=1$ situated below the straight line given by $y-1=0$ (Fig. 102). 3) The half of the ellipse $x^2/16+(y+3)^2/4=1$ situated in the left half-plane (Fig. 103). 4) The half of the ellipse $(x+5)^2/4+ +(y-1)^2/9=1$ situated to the right of the straight line given by $x+5= =0$ (Fig. 104). **473.** 1) $(x-2)^2/169+y^2/25=1$, 2) $2x^2-2xy+2y^2- -3=0$, 3) $68x^2+48xy+82y^2-625=0$, 4) $11x^2+2xy+11y^2-48x-$

$-48y-24=0$. **474.** $5x^2+9y^2+4x-18y-55=0$. **475.** $4x^2+3y^2+32x-14y+59=0$. **476.** $4x^2+5y^2+14x+40y+81=0$. **477.** $7x^2-2xy+7y^2-46x+2y+71=0$. **478.** $17x^2+8xy+23y^2+30x-40y-175=0$. **479.** $x^2+2y^2-6x+24y+31=0$. **480.** (4, 3/2), (3, 2). **481.** (3, 8/5), line touches ellipse. **482.** Line passes outside ellipse. **483.** The straight line 1) intersects, 2) passes outside, 3) touches, the ellipse. **484.** The straight line 1) intersects the ellipse if $|c|<5$, 2) touches it if $c=\pm 5$, 3) passes outside it if $|c|>5$. **485.** $m^2a^2+b^2=c^2$. **486.** $xx_1/a^2+yy_1/b^2=1$. **488.** $3x+2y-10=0$, $3x+2y+10=0$. **489.** $x+y-5=0$, $x+y+5=0$. **490.** $2x-y-12=0$, $2x-y+12=0$. $d=24\sqrt{5}/5$. **491.** $M_1(-3, 2)$, $d=\sqrt{13}$. **492.** $x+y-5=0$, $x+4y-10=0$. **493.** $4x-5y-10=0$. **494.** $d=18$. **495.** $x^2/20+y^2/5=1$, or $x^2/80+4y^2/5=1$. **496.** $x^2/40+y^2/10=1$. **499.** $x^2/17+y^2/8=1$.

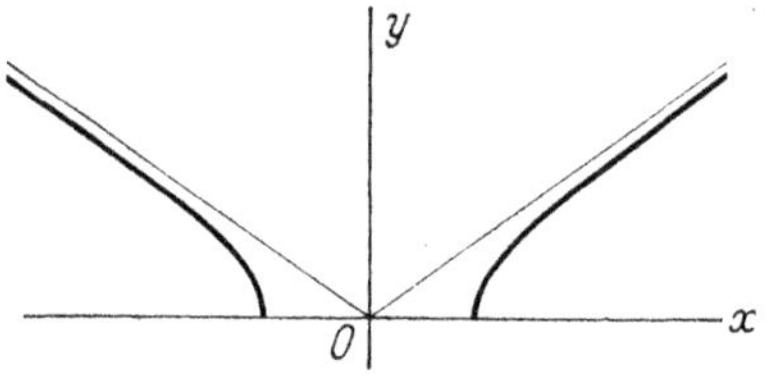

FIG. 105

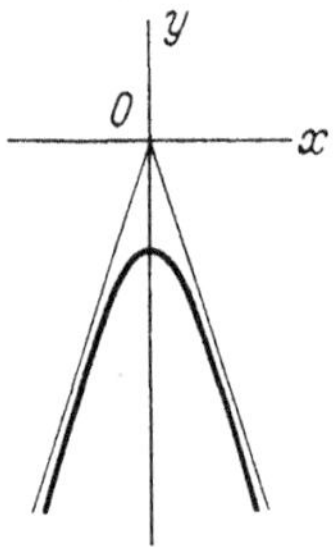

FIG. 106

Hint: Use the result of q.**498.** **500.** $x^2/25+y^2/4=1$. *Hint:* Use the result of q.**498.** **502.** $2x+11y-10=0$. *Hint:* Use the result of q.**501.** **503.** (3, 2), (3, −2). **504.** $R=(mn\sqrt{2})/\sqrt{(m^2+n^2)}$. **505.** 10, $5\sqrt{3}$. **506.** $\varphi=60°$. **507.** 16, 8. **508.** 60°. **509.** Into the ellipse given by $x^2/25+y^2/16=1$. **510.** $x^2+y^2=9$. **511.** $x^2/36+y^2/16=1$. **512.** $q=4/3$. **513.** $q=2/3$. **514.** $q_1=4/3$, $q_2=4/5$. **515.** 1) $x^2/25-y^2/16=1$. 2) $x^2/9-y^2/16=1$. 3) $x^2/4-y^2/5=1$. 4) $x^2/64-y^2/36=1$. 5) $x^2/36-y^2/64=1$. 6) $x^2/144-y^2/25=1$. 7) $x^2/16-y^2/9=1$. 8) $x^2/4-y^2/5=1$. 9) $x^2/64-y^2/36=1$. **516.** 1) $x^2/36-y^2/324=-1$. 2) $x^2/16-y^2/9=-1$. 3) $x^2/100-y^2/576=-1$. 4) $x^2/24-y^2/25=-1$. 5) $x^2/9-y^2/16=-1$. **517.** 1) $a=3$, $b=2$. 2) $a=4$, $b=1$. 3) $a=4$, $b=2$. 4) $a=1$, $b=1$. 5) $a=5/2$, $b=5/3$. 6) $a=1/5$, $b=1/4$, 7) $a=1/3$, $b=1/8$. **518.** 1) $a=3$, $b=4$. 2) $F_1(-5, 0)$, $F_2(5, 0)$. 3) $\varepsilon=5/3$. 4) $y=\pm 4x/3$. 5) $x=\pm 9/5$. **519.** 1) $a=3$, $b=4$. 2) $F_1(0, -5)$, $F_2(0, 5)$. 3) $\varepsilon=5/4$. 4) $y=\pm 4x/3$. 5) $y=\pm 16/5$. **520.** 12 units of area. **521.** 1) The part of the hyperbola given by $x^2/9-y^2/4=1$ situated in the upper half-plane (Fig. 105). 2) The branch of the hyperbola given by $x^2-y^2/9=-1$ lying in the lower half-plane (Fig. 106).

3) The branch of the hyperbola given by $x^2/16-y^2/9=1$ situated in the left half-plane (Fig. 107). 4) The branch of the hyperbola given by $x^2/25-$ $-y^2/4=-1$ situated in the upper half-plane (Fig. 108). **522.** $x+4\sqrt{5}y+$ $+10=0$, $x-10=0$. **523.** $r_1=9/4$, $r_2=41/4$. **524.** 8. **525.** 12.

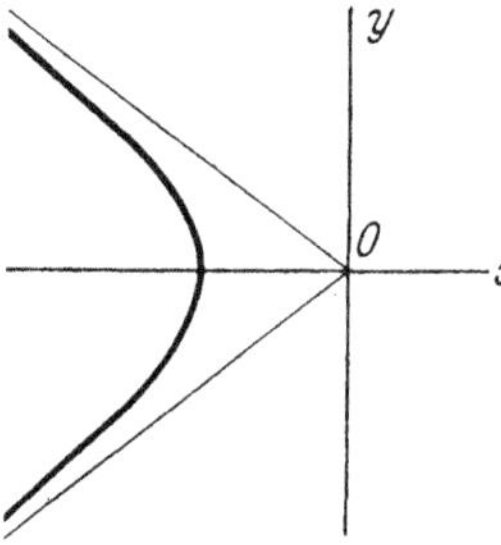

FIG. 107

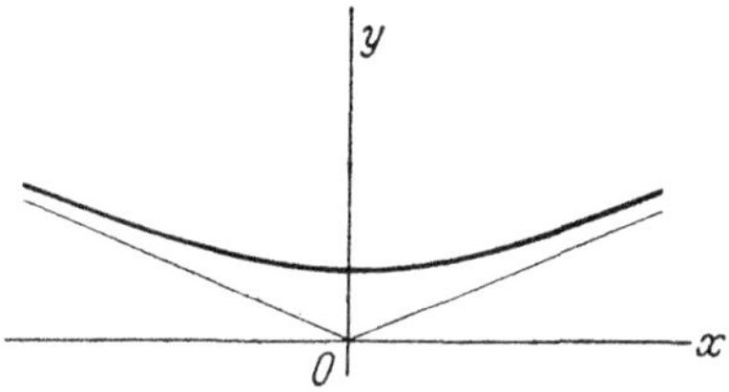

FIG. 108

526. 10. **527.** 27. **528.** (10, 9/2), (10, − 9/2). **529.** $(-6, 4\sqrt{3})$, $(-6$ $-4\sqrt{3})$. **530.** 25/12, 313/12. **531.** See Fig. 109. **532.** 1) $x^2/32-y^2/8=$ $=1$. 2) $x^2-y^2=16$. 3) $x^2/4-y^2/5=1$ or $x^2/(\frac{61}{9})-y^2(\frac{305}{16})=1$. 4) $x^2/18-$ $-y^2/8=1$. 5) $x^2/16-y^2/9=1$. **533.** $\varepsilon=\sqrt{2}$. **534.** $\varepsilon=\sqrt{3}$. **535.** $x^2/4-y^2/12=1$. **536.** $x^2/60-y^2/40=1$. **540.** 1) $(x-x_0)^2/a^2-(y-$

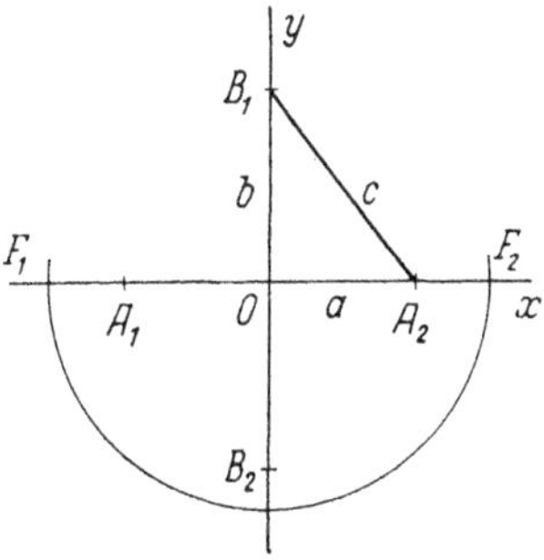

FIG. 109

$-y_0)^2/b^2=1$. 2) $(x-x_0)^2/a^2-(y-y_0)^2/b^2=-1$. **541.** 1) $C(2, -3)$, $a=3$, $b=4$, $\varepsilon=5/3$, equations of directrices $5x-1=0$, $5x-19=0$, equations of asymptotes $4x-3y-17=0$, $4x+3y+1=0$. 2) $C(-5, 1)$, $a=8$, $b=6$, $\varepsilon=5/4$, equations of directrices $x=-11{\cdot}4$, $x=1{\cdot}4$, equations of asymptotes $3x+4y+11=0$, $3x-4y+19=0$. 3) $C(2, -1)$, $a=3$, $b=4$, $\varepsilon=5/4$, equations of directrices $y=-4{\cdot}2$, $y=2{\cdot}2$, equations of asymptotes $4x+3y-5=0$, $4x-3y-11=0$. **542.** 1) The part of the hyperbola given by $(x-2)^2/9-(y+1)^2/4=1$ situated above the straight line given by $y+1=0$ (Fig. 110). 2) The branch of the hyperbola given by $(x-3)^2/4-(y-7)^2/9=-1$ situated below the straight line given by $y-7=0$ (Fig. 111). 3) The branch of the hyperbola given

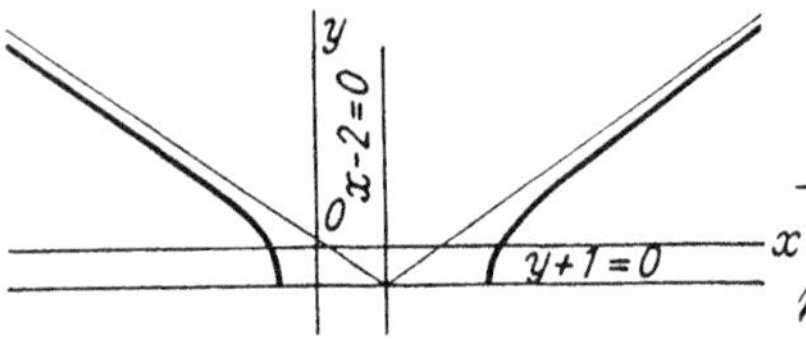

FIG. 110

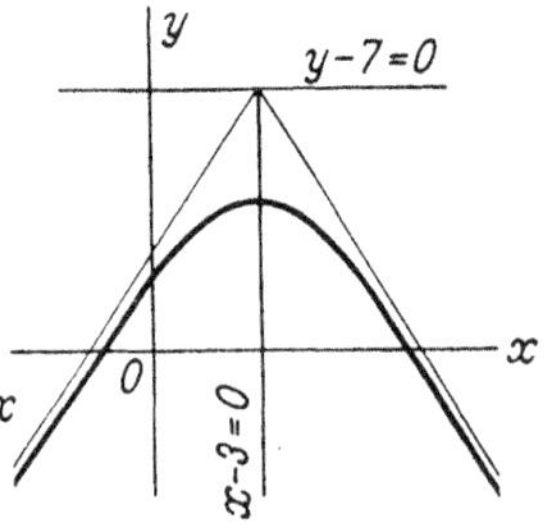

FIG. 111

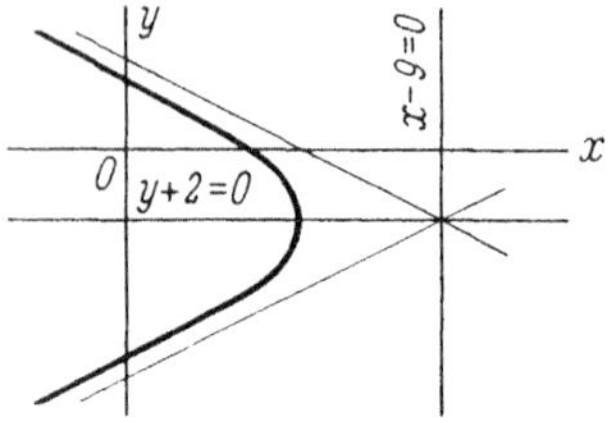

FIG. 112

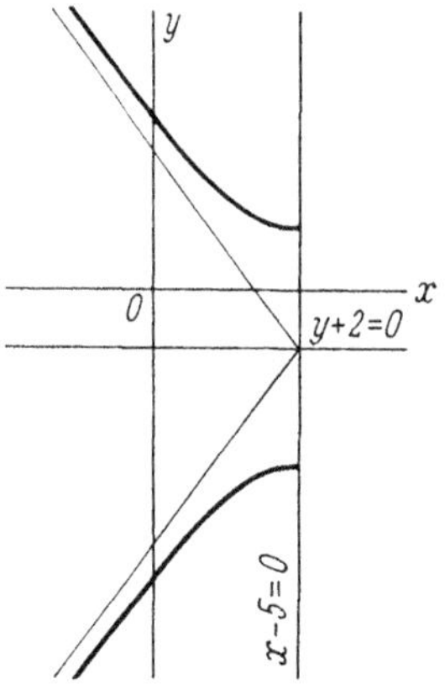

FIG. 113

by $(x-9)^2/16-(y+2)^2/4=1$ situated to the left of the straight line given by $x-9=0$ (Fig. 112). 4) The part of the hyperbola given by $(x-5)^2/9-$ $-(y+2)^2/16=-1$ situated to the left of the straight line given by $x-5=$ $=0$ (Fig. 113). **543.** 1) $(x-3)^2/144-(y-2)^2/25=1$. 2) $24xy+7y^2-$ $-144=0$. 3) $2xy+2x-2y+7=0$. **544.** $x^2/16-y^2/9=1$. **545.** $x^2/25-y^2/144=-1$. **546.** $x^2-4y^2-6x-24y-47=0$. **547.** $7x^2-6xy-$ $-y^2+26x-18y-17=0$. **548.** $91x^2-100xy+16y^2-136x+86y-47=$

$= 0$. **549.** $xy = a^2/2$ if the old axes are rotated through $-45°$; $xy = -a^2/2$ if the old axes are rotated through $+45°$. **550.** 1) $C(0, 0)$, $a = b = 6$, equations of asymptotes $x = 0$, $y = 0$. 2) $C(0, 0)$, $a = b = 3$, equations of asymptotes $x = 0$, $y = 0$. 3) $C(0, 0)$, $a = b = 5$, equations of asymptotes $x = 0$, $y = 0$. **551.** $(6, 2)$, $(14/3, -2/3)$. **552.** $(25/4, 3)$: the straight line touches the hyperbola. **553.** The straight line passes outside the hyperbola. **554.** 1) Touches. 2) Intersects hyperbola in two points. 3) Passes outside hyperbola. **555.** 1) Intersects hyperbola if

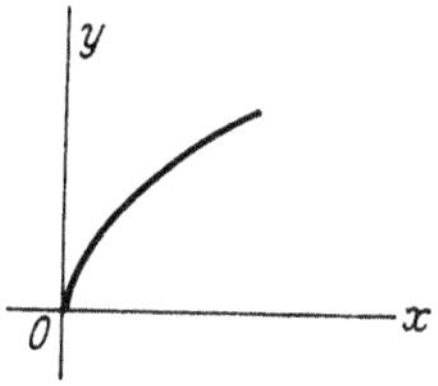

FIG. 114

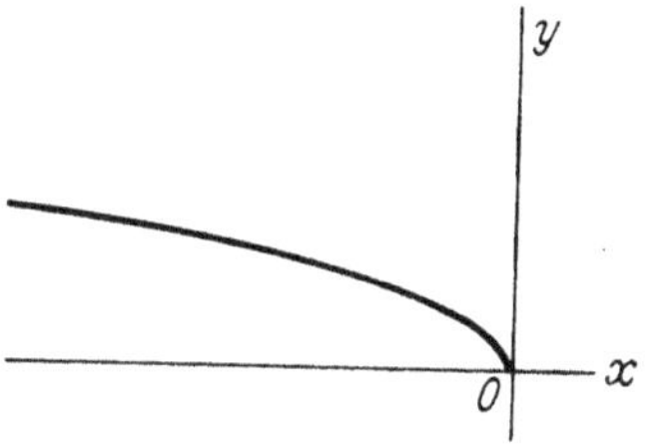

FIG. 115

$|c| > 4{\cdot}5$. 2) Touches the hyperbola if $c = \pm 4{\cdot}5$. 3) Passes outside hyperbola if $|c| < 4{\cdot}5$. **556.** $m^2a^2 - b^2 = c^2$. **557.** $x_1x/a^2 - y_1y/b^2 = 1$. **559.** $3x - 4y - 10 = 0$, $3x - 4y + 10 = 0$. **560.** $10x - 3y - 32 = 0$, $10x - 3y + 32 = 0$. **561.** $x + 2y - 4 = 0$, $x + 2y + 4 = 0$, $d = 8\sqrt{5}/5$. **562.** $M_1(-6, 3)$, $d = 11\sqrt{13}/13$. **563.** $5x - 3y - 16 = 0$, $13x + 5y + 48 = 0$. **564.** $2x + 5y - 16 = 0$. **565.** $d = 17\sqrt{10}/10$. **566.** $x^2/5 - y^2/45 = 1$, $3x^2/10 - 4y^2/45 = 1$. **567.** $x^2/16 - y^2/4 = 1$. **568.** $x = -4$, $x = 4$, $y = -1$, $y = 1$. **572.** $x^2/5 - y^2/4 = 1$. **573.** $x^2/16 - y^2/9 = 1$. **575.** $2x + 11y + 6 = 0$. *Hint:* Use the result of q.**574.** **577.** $x^2 - y^2 = 16$. **578.** $x^2/16 - y^2/9 = 1$. **579.** $x^2/25 - y^2/4 = 1$. **580.** $q = 2/3$. **581.** $q = 2$. **582.** $q_1 = 2$, $q_2 = 5/7$. **583.** 1) $y^2 = 6x$, 2) $y^2 = -x$, 3) $x^2 = y/2$, 4) $x^2 = -6y$. **584.** 1) $2a = 3$: in the right half-plane symmetrically about the x-axis. 2) $2a = 2{\cdot}5$: in the upper half-plane symmetrically about the y-axis. 3) $2a = 2$: in the left half-plane symmetrically about the x-axis. 4) $2a = 1/2$: in the lower half-plane symmetrically about the y-axis. **585.** 1) $y^2 = 4x$, 2) $y^2 = -9x$, 3) $x^2 = y$, 4) $x^2 = -2y$. **586.** 40 cm. **587.** $x^2 = -12y$. **588.** 1) The part of the parabola given by $y^2 = 4x$ situated in the 1st quadrant (Fig. 114). 2) The part of the parabola given by $y^2 = -x$ situated in the 2nd quadrant (Fig. 115). 3) The part of the parabola given by $y^2 = -18x$ situated in the 3rd quadrant (Fig. 116). 4) The part of the parabola given by $y^2 = 4x$ situated in the 4th quadrant (Fig. 117). 5) The part of the parabola given by $x^2 = 5y$ situated in the 1st quadrant. (Fig. 118). 6) The part of the parabola given by $x^2 = -25y$ situated in the 3rd quadrant (Fig. 119). 7) The part of the parabola given by $x^2 = 3y$

situated in the 2nd quadrant (Fig. 120). 8) The part of the parabola given by $x^2 = -16y$ situated in the 4th quadrant (Fig. 121). **589.** $F(6, 0)$, $x+6 = 0$. **590.** 12. **591.** 6. **592.** $(9, 12)$, $(9, -12)$. **593.** $y^2 = -28x$.

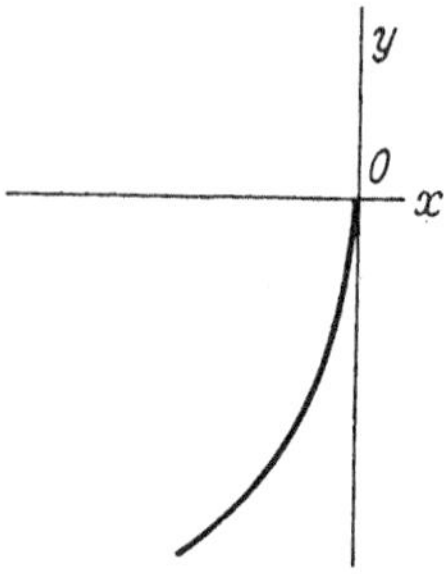

FIG. 116

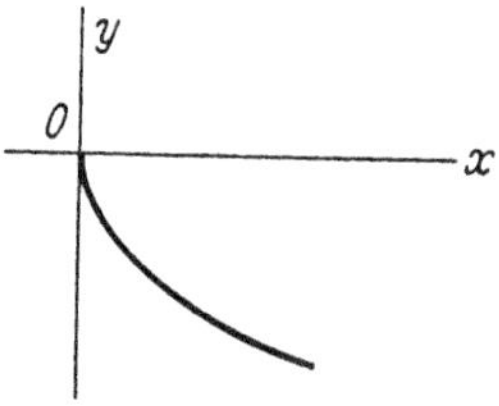

FIG. 117

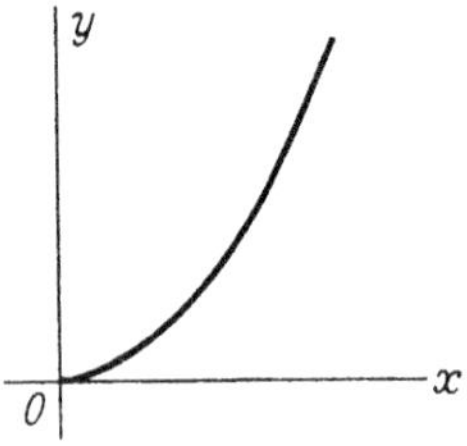

FIG. 118

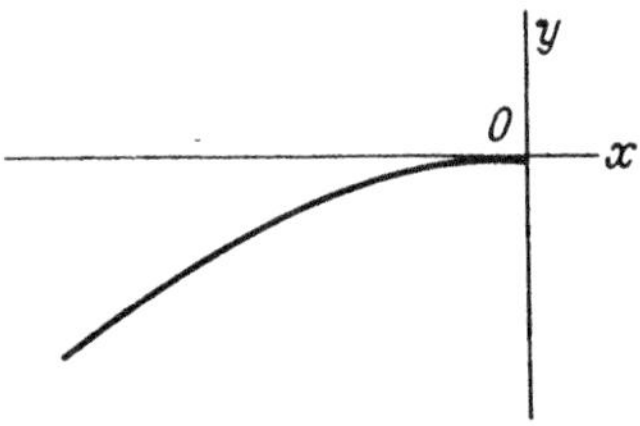

FIG. 119

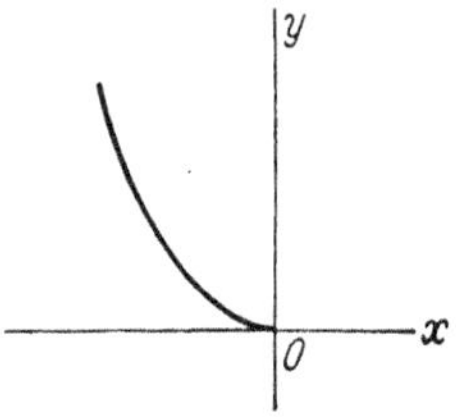

FIG. 120

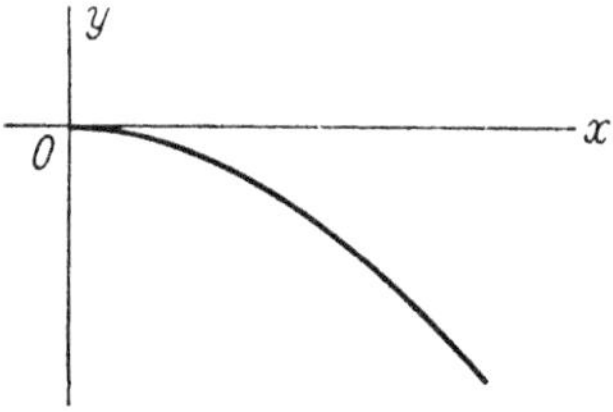

FIG. 121

594. 1) $(y-\beta)^2 = 4a(x-\alpha)$. 2) $(y-\beta)^2 = -4a(x-\alpha)$. **595.** 1) $(x-\alpha)^2 = 4a(y-\beta)$, 2) $(x-\alpha)^2 = -4a(y-\beta)$. **596.** 1) $A(2, 0)$, $2a = 2$, $x-1 = 0$. 2) $A(2/3, 0)$, $2a = 3$, $6x-13 = 0$. 3) $A(0, -1/3)$, $2a = 3$, $6y+11 = 0$. 4) $A(0, 2)$, $2a = 1/2$, $4y-9 = 0$. **597.** 1) $A(-2, 1)$, $2a = 2$. 2) $A(1, 3)$,

$2a = 1/8$. 3) $A(6, -1)$, $2a = 3$. **598.** 1) $A(-4, 3)$, $2a = 1/4$. 2) $A(1, 2)$, $2a = 2$. 3) $A(0, 1)$. $2a = 1/2$. **599.** 1) The part of the parabola given by $(y-3)^2 = 16(x-1)$ situated below the straight line given by $y-3 = 0$ (Fig. 122). 2) The part of the parabola given by $(x+4)^2 = 9(y+5)$ situated to the right of the straight line given by $x+4 = 0$ (Fig. 123). 3) The part of the parabola given by $(x-2)^2 = -2(y-3)$ situated to the left

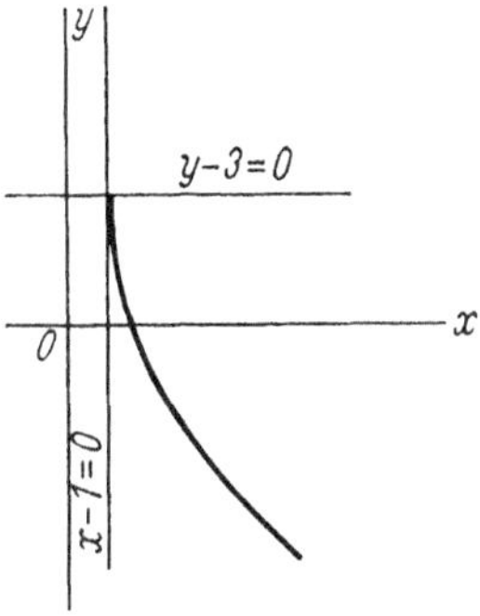

FIG. 122

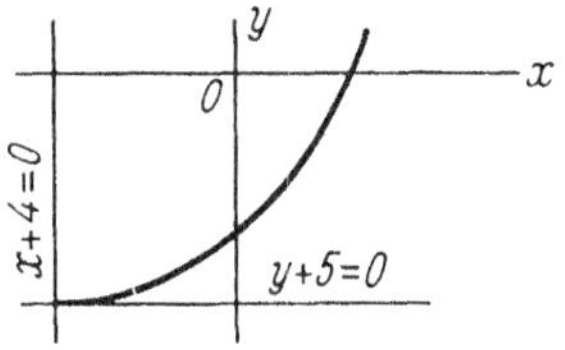

FIG. 123

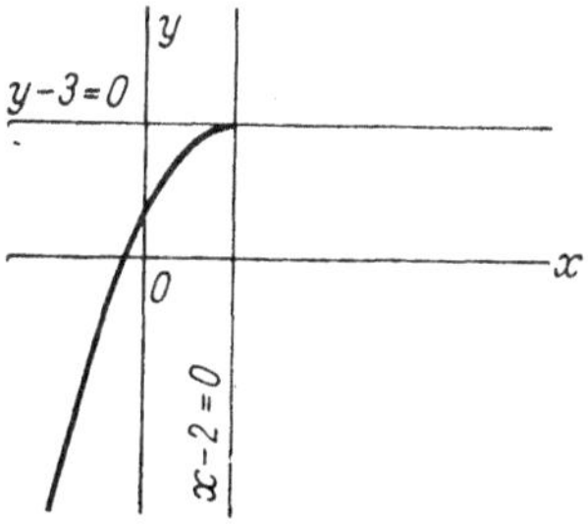

FIG. 124

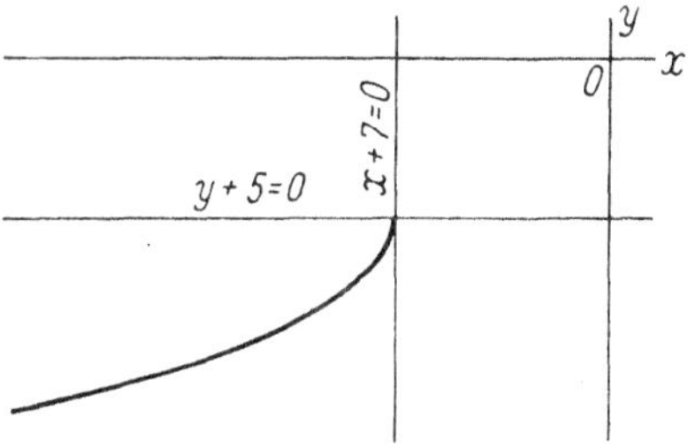

FIG. 125

of the straight line given by $x-2 = 0$ (Fig. 124). 4) The part of the parabola given by $(y+5)^2 = -3(x+7)$ situated below the straight line given by $y+5 = 0$ (Fig. 125). **600.** $x = y^2/4 - y + 7$. **601.** $y = x^2/8 - x + 3$. **602.** $x^2 + 2xy + y^2 - 6x + 2y + 9 = 0$. **603.** $F(9, -8)$. **604.** $4x^2 - 4xy + y^2 + 32x + 34y + 89 = 0$. **605.** $(2, 1)$, $(-6, 9)$. **606.** $(-4, 6)$: the straight line touches the parabola. **607.** The straight line and parabola do not intersect. **608.** 1) Touches the parabola. 2) Intersects the parabola in two points. 3) Passes outside the parabola. **609.** 1) $n < 1/2$, 2) $m = 1/2$, 3) $m > 1/2$. **610.** $a = mc$. **612.** $y_1 y = 2a(x + x_1)$. **613.** $x + y + 2 = 0$. **614.** $2x - y - 16 = 0$. **615.** $d = 2\sqrt{13}$. **616.** $M_1(9, -24)$,

$d = 10$. **617.** $3x-y+3=0$ and $3x-2y+12=0$. **619.** $5x-18y+$ $+25=0$. **620.** $d=174/13$. **621.** $(6, 12)$, $(6, -12)$. **622.** $(10, \sqrt{30})$, $(10, -\sqrt{30})$, $(2, \sqrt{6})$, $(2, -\sqrt{6})$. **623.** $(2, 1)$, $(-1, 4)$, $\left(\frac{3+\sqrt{13}}{2}, \frac{7+\sqrt{13}}{2}\right)$, $\left(\frac{3-\sqrt{13}}{2}, \frac{7-\sqrt{13}}{2}\right)$. **625.** $y-18=0$. *Hint:* use the result of q. **624.** **628.** 1) $\varrho = \frac{16}{5-3\cos\theta}$, 2) $\varrho = \frac{16}{5+3\cos\theta}$. **629.** 1) $\theta = \frac{9}{4-5\cos\theta}$, 2) $\varrho = -\frac{9}{4-5\cos\theta}$. **639.** 1) $\varrho = \frac{144}{5+13\cos\theta}$, 2) $\varrho = -\frac{144}{5+13\cos\theta}$.

631. $\varrho = \frac{3}{1-\cos\theta}$. **632.** 1) Ellipse. 2) Parabola. 3) Branch of hyperbola. 4) Ellipse. 5) Branch of hyperbola. 6) Parabola. **633.** 13, 12. **634.** 8, 6. **635.** $\varrho = -21/(2\cos\theta)$, $\varrho = 29/(2\cos\theta)$. **636.** Equations of directrices: $\varrho = -34/(5\cos\theta)$, $\varrho = 16/(5\cos\theta)$. Equations of asymptotes: $\varrho = \frac{20}{3\sin\theta - 4\cos\theta}$, $\varrho = -\frac{20}{3\sin\theta+4\cos\theta}$. **637.** $(6, \pi/4)$, $(6, -\pi/4)$. **638.** $(3, 2\pi/3)$, $(3, -2\pi/3)$. **639.** 1) $(p/2, \pi)$, 2) $(p, \pi/2)$, $(p, -\pi/2)$. **640.** $\varrho^2 = -\frac{-b^2}{\varepsilon^2\cos^2\theta-1}$. **641.** $\varrho^2 = \frac{b^2}{\varepsilon^2\cos^2\theta-1}$.

642. $\varrho = \frac{2p\cos\theta}{\sin^2\theta}$. **643.** $8x+25y=0$, **644.** $9x-32y-73=0$. **645.** $x-$ $-y=0$, $x+4y=0$. **646.** $x+2y=0$, $8x-9y=0$. **647.** $x+2y=0$, $2x-3y=0$. **654.** $2x-5y=0$. **655.** $7x+y-20=0$. **656.** $x-8y=0$, $2x-y=0$. **657.** $x-2y=0$, $3x-y=0$; $x+2y=0$, $3x+y=0$. **661.** $y+2=0$. **662.** $2x-y+1=0$. **665.** 1), 2), 5), 8) have a unique centre, 3), 7) have no centre, 4), 6) have infinitely many centres. **666.** 1) $(3, -2)$, 2) $(0, -5)$, 3) $(0, 0)$, 4) $(-1, 3)$. **667.** 1) $x-3y-6=0$, 2) $2x+y-2=0$, 3) $5x-y+4=0$. **668.** 1) $9x^2-18xy+6y^2+2=0$, 2) $6x^2+4xy+y^2-$ $-7=0$, 3) $4x^2+6xy+y^2-5=0$, 4) $4x^2+2xy+6y^2+1=0$. **669.** a) $m \neq 4$, *n arbitrary*; b) $m=4$, $n \neq 6$; c) $m=4$, $n=6$. **670.** a) $m=$ $=2$, b) $m_1=-1$, $m_2=5$, c) for any $m \neq 2$ satisfying $-1<m<5$, d) for $m<-1$ and for $m>5$. **671.** $x^2-8y^2-4=0$. **672.** x^2+xy+ $+y^2+3y=0$. **673.** 1) Elliptic equation; defines ellipse given by $\tilde{x}^2/9+$ $+\tilde{y}^2/4=1$ in the new coordinates; new origin $\tilde{O}(5, -2)$. 2) Hyperbolic; $\tilde{x}^2/16-\tilde{y}^2/9=1$; $\tilde{O}(3, -2)$. 3) Elliptic; $\tilde{x}^2/4+\tilde{y}^2/9=-1$ having no geometrical representation ("imaginary ellipse"). 4) Hyperbolic; $4\tilde{x}^2-$ $-\tilde{y}^2=0$, defining a degenerate hyperbola (pair of intersecting straight lines); $\tilde{O}(-1, -1)$. 5) Elliptic; $2\tilde{x}^2+3\tilde{y}^2=0$, defining a degenerate ellipse (single point). **674.** 1) Hyperbolic equation: $\tilde{x}'^2/9-\tilde{y}'^2/4=1$, defining a hyperbola: $\tan\alpha=-2$, $\cos\alpha=1/\sqrt{5}$, $\sin\alpha=-2/\sqrt{5}$. 2) Elliptic equation: $\tilde{x}'^2/16+\tilde{y}'^2/4=1$: $\alpha=45°$. 3) Elliptic equation: $\tilde{x}'^2+4\tilde{y}'^2=0$, defining a degenerate ellipse (a single point): $\tan\alpha=2$, $\cos\alpha=1/\sqrt{5}$,

$\sin\alpha = 2/\sqrt{5}$. 4) Hyperbolic equation: $\tilde{x}'^2 - \tilde{y}'^2 = 0$, defining a degenerate hyperbola (a pair of intersecting straight lines): $\tan\alpha = 2/3$, $\cos\alpha = 3/\sqrt{13}$, $\sin\alpha = 2/\sqrt{13}$. 5) Elliptic equation: $\tilde{x}'^2/4 + \tilde{y}'^2 = -1$, defining an "imaginary ellipse": $\alpha = 45°$. **675.** 1) hyperbolic, 2) elliptic, 3) parabolic, 4) elliptic, 5) parabolic, 6) hyperbolic. **676.** 1) Hyperbolic equation: defines a hyperbola, whose equation takes the form $\tilde{x}'^2 - \tilde{y}'^2/4 = 1$ after the two transformations of coordinates $x = \tilde{x}+2$, $y = \tilde{y}-1$ and

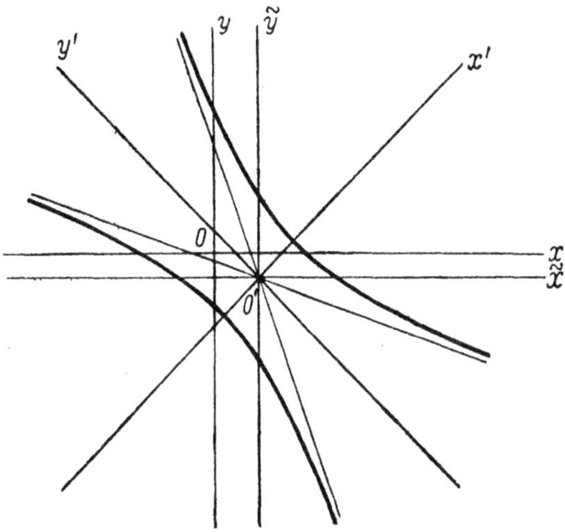

FIG. 126

$\tilde{x} = (\tilde{x}' - \tilde{y}')/\sqrt{2}$, $\tilde{y} = (\tilde{x}' + \tilde{y}')/\sqrt{2}$ (Fig. 126). 2) Elliptic equation: defines an ellipse whose equation takes the form $\tilde{x}'^2/16 + \tilde{y}'^2/9 = 1$ after the two successive transformations of coordinates $x = \tilde{x}-1$, $y = \tilde{y}+1$ and $\tilde{x} = (\tilde{x}' - \tilde{y}')/\sqrt{2}$, $\tilde{y} = (\tilde{x}' + \tilde{y}')/\sqrt{2}$ (Fig. 127). 3) Hyperbolic equation: defines a hyperbola whose equation takes the form $\tilde{x}'^2/9 - \tilde{y}'^2/36 = 1$ after the two successive transformations of coordinates $x = \tilde{x}+3$, $y = \tilde{y}-4$ and $\tilde{x} = (\tilde{x}' - 2\tilde{y}')/\sqrt{5}$, $\tilde{y} = (2\tilde{x}' + \tilde{y}')/\sqrt{5}$ (Fig. 128). 4) Hyperbolic equation: defines a degenerate hyperbola (a pair of intersecting straight lines) whose equation takes the form $\tilde{x}'^2 - 4\tilde{y}'^2 = 0$ after the two successive transformations of coordinates $x = \tilde{x}-2$, $y = \tilde{y}$ and $\tilde{x} = (\tilde{x}' + 3\tilde{y}')/\sqrt{10}$, $\tilde{y} = (-3\tilde{x}' + \tilde{y}')/\sqrt{10}$ (Fig. 129). 5) Elliptic equation: defines an "imaginary ellipse" whose equation takes the form $\tilde{x}'^2 + 2\tilde{y}'^2 = -1$ after the two successive transformations of coordinates $x = \tilde{x}-1$, $y = \tilde{y}$ and $\tilde{x} = (\tilde{x}' + 3\tilde{y}')/\sqrt{10}$, $\tilde{y} = (-3\tilde{x}' + \tilde{y}')/\sqrt{10}$. 6) Elliptic equation: defines a degenerate ellipse (a single point) whose equation takes the form $2\tilde{x}'^2 + 3\tilde{y}'^2 = 0$ after the two successive transformations of coordinates

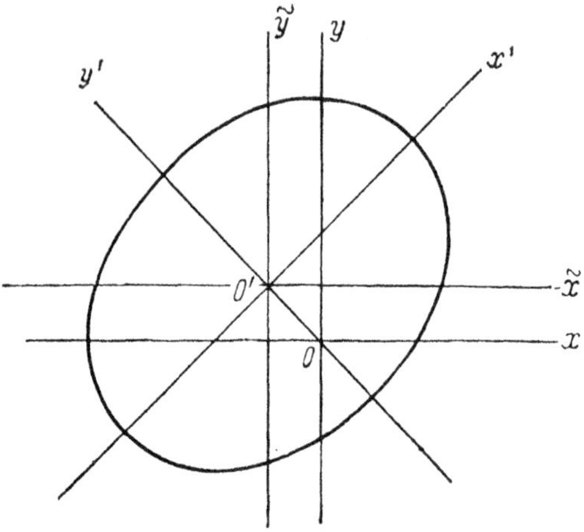

FIG. 127

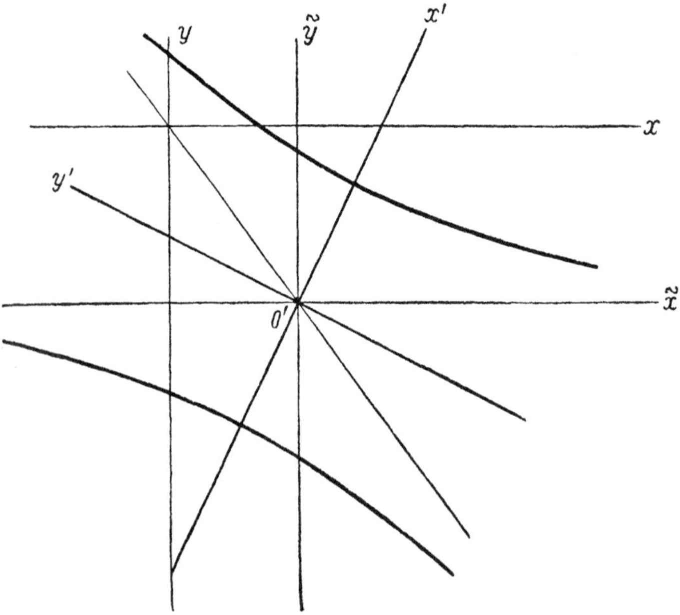

FIG. 128

$x = \tilde{x}$, $y = \tilde{y} - 2$ and $\tilde{x} = (\tilde{x}' - \tilde{y}')/\sqrt{2}$, $\tilde{y} = (\tilde{x}' + \tilde{y}')/\sqrt{2}$. **677.** 1) $x^2/30 + y^2/5 = 1$, ellipse. 2) $9x^2 - 16y^2 = 5$, hyperbola. 3) $x^2 - 4y^2 = 0$, degenerate hyperbola (pair of intersecting straight lines given by $x - 2y = 0$,

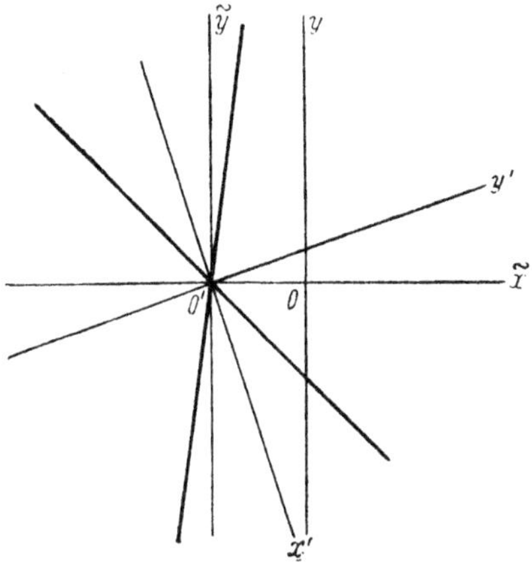

FIG. 129

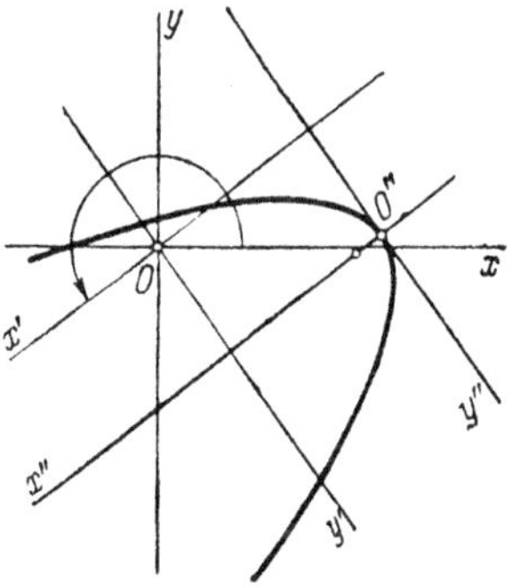

FIG. 130

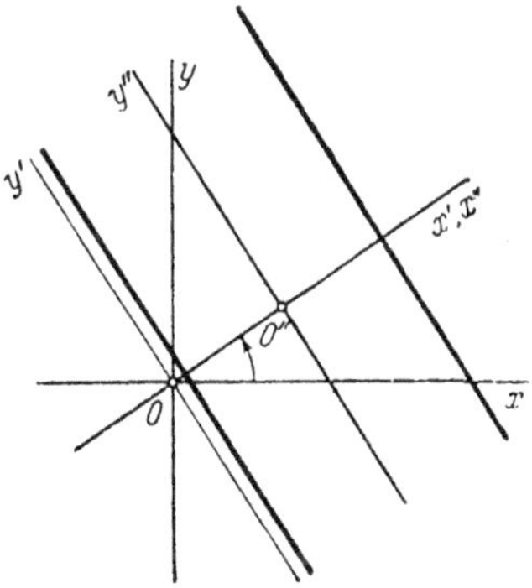

FIG. 131

$x + 2y = 0$). 4) $2x^2 + 3y^2 = -1$, "imaginary ellipse". 5) $x^2 + 2y^2 = 0$, degenerate ellipse (a single point, the origin). 6) $x^2/9 + y^2/4 = 1$, ellipse. 7) $x^2/4 - y^2 = 1$, hyperbola. 8) $x^2/9 + y^2 = 1$, ellipse. **678.** 1) 3 and 1, 2) 3 and 2, 3) 1 and 1/2, 4) 3 and 2. **679.** a) $x = 2$, $y = 3$, b) $x = 3$, $y = -3$, c) $x = 1$, $y = -1$, d) $x = -2$, $y = 1$. **680.** 1) 2 and 1, 2) 5 and 1,

3) 4 and 2, 4) 1 and 1/2. **681.** a) $x+y-1=0$, $3x+y+1=0$. b) $x-4y-2=0$, $x-2y+2=0$. c) $x-y=0$, $x-3y=0$, d) $x \quad y-3=0$, $x+3y-3=0$. **682.** 1) Ellipse. 2) hyperbola. 3) degenerate hyperbola (pair of intersecting straight lines). 4) "imaginary ellipse". 5) degenerate ellipse (a single point). **689.** 1) Parabolic equation: defines a parabola whose equation takes the form $\tilde{y}'^2 = 2\tilde{x}'$ after the two successive transformations $x=(-4x'+3y')/5$, $y=(-3x'-4y')/5$ and $x'=\tilde{x}'-3$, $y'=\tilde{y}'+2$ (Fig. 130). 2) Parabolic equation: defines a degenerate parabola (a pair of parallel straight lines) whose equation takes the form $\tilde{x}'^2 = 1$ after the two successive transformations of coordinates $x=(3x'-2y')/\sqrt{13}$, $y=(2x'+3y')/\sqrt{13}$ and $x'=\tilde{x}'+4/\sqrt{13}$, $y'=\tilde{y}'$ (Fig. 131). 3) Parabolic equation: has no geometrical representation: equation can be put into the form $\tilde{y}'^2+1=0$ by two successive transformations of coordinates $x=(3x'-4y')/5$, $y=(4x'+3y')/5$ and $x'=\tilde{x}'$, $y'=\tilde{y}'-4$. **690.** 1) $y^2=6x$, parabola. 2) $y^2=25$, degenerate parabola (a pair of parallel straight lines given by $y-5=0$, $y+5=0$). 3) $y^2=0$, degenerate parabola (pair of coincident straight lines, coinciding with the x-axis). **693.** 1) $(x+2y)^2+4x+y-15=0$, 2) $(3x-y)^2-x+2y-14=0$, 3) $(5x-2y)^2+3x-y+11=0$, 4) $(4x+2y)^2-5x+7y=0$, 5) $(3x-7y)^2+3x-2y-24=0$. **697.** 1) 3, 2) 3, 3) $\sqrt{2}$, 4) $\frac{2}{3}10$. **699.** a) $2x+y-5=0$, $2x+y-1=0$. b) $2x-3y-1=0$, $2x-3y+11=0$. c) $5x-y-3=0$, $5x-y+5=0$. **700.** a) $x-3y+2=0$, b) $3x+5y+7=0$, c) $4x-2y-9=0$. **701.** $(x^2+y^2)^2-2c^2(x^2-y^2)=a^4-c^4$. **702.** $(x^2+y^2)^2=2a^2(x^2-y^2)$. $\varrho^2=2a^2\cos 2\theta$. **703.** $\varrho^2=S\sin 2\theta$. $(x^2+y^2)^2=2Sxy$. **705.** $\varrho=v\theta/\omega$, $\varrho=-v\theta/\omega$. **706.** $(2r-x)y^2=x^3$. **707.** $x(a^2+y^2)=a^3$. **708.** $\varrho=\dfrac{a}{\cos\theta}\pm b$. $x^2y^2+(x+a)^2(x^2-b^2)=0$. **709.** $\varrho=\dfrac{a}{\cos\theta}\pm a\tan\theta$. $x^2[(x+a)^2+y^2]=a^2y^2$. **710.** $\varrho=2a\cos\theta\pm b$. $(x^2+y^2-2ax)^2=b^2(x^2+y^2)$. **711.** $\varrho=a|\sin 2\theta|$. $(x^2+y^2)^3=4a^2x^2y^2$. **712.** $x=a\cos^3 t$, $y=a\sin^3 t$. $x^{\frac{2}{3}}+y^{\frac{2}{3}}=a^{\frac{2}{3}}$ **713.** $\varrho=a\cos^3\theta$. $(x^2+y^2)^2=ax^3$. **714.** $x=a(\cos t+t\sin t)$, $y=a(\sin t-t\cos t)$. **715.** $x=a(t-\sin t)$, $y=a(1-\cos t)$, $x+\sqrt{[y(2a-y)]}=a\cos^{-1}\dfrac{a-y}{a}$. **716.** $x=a(2\cos t-\cos 2t)$, $y=a(2\sin t-\sin 2t)$. $\varrho=2a(1-\cos\theta)$. **717.** $x=(a+b)\cos t-a\cos\dfrac{a+b}{a}t$, $y=(a+b)\sin t-a\sin\dfrac{a+b}{a}t$. **718.** $x=(b-a)\cos t+a\cos\dfrac{b-a}{a}t$, $y=(b-a)\sin t-a\sin\dfrac{b-a}{a}t$.

MADE IN GREAT BRITAIN

Printed in Dunstable, United Kingdom

91221473R20111